STUDENT STUDY GUIDE AND WORKBOOK

for Essentials of Exercise Physiology

SECOND EDITION

STUDENT STUDY GUIDE AND WORKBOOK

for Essentials of Exercise Physiology

SECOND EDITION

- **Victor L. Katch**
  Professor of Movement Science
  Division of Kinesiology
  Associate Professor of Pediatrics
  School of Medicine
  University of Michigan
  Ann Arbor, Michigan

- **Frank I. Katch**
  Professor of Exercise Science
  University of Massachusetts
  Amherst, Massachusetts

- **William D. McArdle**
  Professor of Physical Education
  Queens College
  City University of New York
  Queens, New York

LIPPINCOTT WILLIAMS & WILKINS
A **Wolters Kluwer** Company
Philadelphia · Baltimore · New York · London
Buenos Aires · Hong Kong · Sydney · Tokyo

Editor: Eric Johnson
Managing Editor: Karen Gulliver
Production Manager: Susan Rockwell
Design Coordinator: Mario Fernandez

Copyright © 2000 Lippincott Williams & Wilkins

351 West Camden Street
Baltimore, Maryland 21201-2436 USA

Printed in the United States of America

ISBN 0-7817-2914-9

Visit Lippincott Williams & Wilkins on the Internet: http:/lww.com

03 04
4 5 6 7 8 9 10

Preface

The *Student Study Guide and Workbook* serves as a resource companion to *Essentials of Exercise Physiology, Second Edition.* We designed the *Student Study Guide and Workbook* with two main purposes in mind: (1) to help students better understand the text content by focusing on key terms and concepts, and (2) to organize the material in such a way so students could improve their ability to answer questions from each section within a chapter. The use of bolded words and terms within the text and the workbook allows students to generate a glossary of terms used throughout the text. We have also added sample quizzes (true/false and multiple choice) and crossword puzzles to help reinforce key terms and concepts.

During the past six years, feedback has been encouraging from many instructors and students who have used the workbook with the first edition of *Essential of Exercise Physiology.* **The feedback clearly resonates: Students who use the workbook receive higher grades and gain greater understanding than those who do not use the workbook.** This reinforces our own classroom experiences and evaluations since we first began to use the workbook in 1992. Using the workbook improves student performance and results in higher course and teacher evaluations. In one evaluation study we conducted, some students elected to use the workbook and earn extra-credit points for each workbook chapter they completed. Other students elected to do different extra-credit assignments and earn the same extra-credit points. The results were unequivocal: Students who completed each of the workbook assignments earned on average, one letter grade higher in the course! Were we surprised? No, not really. We were confident students would meet high expectations. And that's how the students also felt. They reported feeling more confident about the text and lecture material and strongly expressed the view they were better prepared for exams. The "workbook students" gave the course instructor higher evaluations, and stated that the course was "more valuable," "better structured," "less stressful" and they felt "more involved" than students not using the workbook. This latter point should not be overlooked. Students who perceive they are more involved in a course, (and who feel better prepared for exams), stand a much greater likelihood for succeeding, with a more positive experience than students without these perceptions.

While students employ different learning strategies, each method still involves the ability to internalize (understand) course material. Our experience indicates that the most common study strategy employs reading and underlining text material; another includes outlining or writing key concepts from the reading, often within the margin of the textbook itself. This workbook enables the student to use each of these techniques; in addition, the workbook directs attention to important key terms and concepts within each chapter section, thus providing additional direction on what needs to be internalized.

The workbook has four sections. The first section asks for definitions for key terms and concepts. Also included are study questions for each major topic heading, a sample quiz, and a crossword puzzle. The study questions follow the sequence of the content material in each chapter. This means it should be easy to locate the answers to each of the questions by referring back to the appropriate section heading within a chapter. In our experience, crossword puzzles always provide an entertaining yet challenging way to study the material.

Section II includes practical self-assessment tests. Assessment of nutrition, physical fitness, and overall health and well being tend to personalize the course material and increase its relevancy. We encourage students to use these assessments and think carefully about how to apply the results to enhance health and wellness. The assessment tools bring to life some of the text material, and show the practical side of using the different techniques and methods.

Section III provides answers to the chapter quizzes and crossword puzzles.

Section IV of the *Student Study Guide and Workbook* presents a list of the nutritive value (caloric content, protein, total fat, carbohydrate, calcium, iron, vitamin B_1, B_2, fiber, and cholesterol) of common foods grouped into 16 food categories. We also include a list of the energy cost values for different physical activities. The ready access to these materials should help students complete several of the practical assessment tests in Section II.

We wish to thank all of the students who provided constructive criticism and ideas on how to make the workbook more useful and interesting. If you have ideas for improving any of our materials, we want to hear from you. Please let us know of additional materials you would like to see included in future editions. Please e-mail any one of us at the addresses listed below. Good luck in your studies, and we hope this course is one of the best you've ever taken!

VICTOR KATCH · Ann Arbor, MI
<vkass@umich.edu>

FRANK I. KATCH · Amherst, MA
<fkatch@exec.sci.umass.edu>

WILLIAM D. MCARDLE · Sound Beach, NY
<kbmwdm@earthlink.net>

Contents

Section 1

- Define Key Terms and Concepts
- Study Questions
- Practice Quizzes
- Crossword Puzzles

Origins of Exercise Physiology: Foundations for the Field of Study

1. AAHPERD

2. Academic discipline

3. ACSM

4. Applied research

5. Archibald Vivian Hill

6. August Krogh

7. Austin Flint, Jr., M.D.

8. Basic research

9. Bengt Saltin

10. Casual relationships

11. Causal relationships

12. Construct

13. Continuous variables

14. Control condition

15. David Bruce Dill

16. De arte Gymnastica apud ancientes

17. Dependent variable

18. Discrete variables

19. Double-blind procedure

20. Empirical research

21. Experimental studies

22. External criticism

23. Fact finding

24. Fernand Lagrange's, "The Physiology of Bodily Exercise"

25. Field studies

26. First exercise physiology research laboratory

27. Galen

28. George Wells Fitz.

29. Harvard Fatigue Laboratory

30. Hawthorne effect

31. Hippocrates

32. Independent variable

33. Individual differences

34. Internal criticism

35. Interval variables

36. Intra-individual variability

37. Laws

38. Medicine and Science in Sports and Exercise

39. Notion of disproof

40. Operational Definitions

41. Ordinal variables

42. Per-Olof Åstrand

43. Placebo effect

44. Primary sources

45. Ratio variables

46. Secondary sources

47. Technological measurement error

48. The Hitchcock's

49. Theoretical research

50. Theory

51. Thomas K. Cureton

52. Variable

Study Questions

INTRODUCTION

List the three components of exercise physiology as an academic discipline

1.

2.

3.

List two fields of study closely aligned to exercise physiology.

1.

2.

ORIGINS OF EXERCISE PHYSIOLOGY: FROM ANCIENT GREECE TO THE UNITED STATES, CIRCA 1850

Earliest Development

Name of the Greek physician-athlete who many consider the "father of preventive medicine."

Indicate the importance of the 1539 text *De arte Gymnastica Apud Ancientes* by the Italian physician Mercurialis.

Early United States Experience

Name the first US medical school.

List two "hot" topics of interest to medicine and exercise physiology in the early 19th century.

1.

2.

Austin Flint, Jr., M.D.: American Physician-Physiologist

Give two reasons for Austin Flint's importance in the history of exercise physiology.

1.

2.

Amherst College Connection

Name the father and son pioneer sport science professors.

1.

2.

Anthropometric Assessment of Body Build

Early work in the area of anthropometry took place at what New England College?

George Wells Fitz, M.D.: A Major Influence

List two reasons for Austin Flint's importance in the history of exercise physiology.

1.

2.

Prelude to Exercise Science: Harvard's Department of Anatomy, Physiology, and Physical Training (B.S. Degree, 1891-1898)

List one unique aspect of the academic major in Harvard's Department of Anatomy, Physiology, and Physical Training.

List three objectives of Harvard's physical education major and exercise physiology research laboratory.

1.

2.

3.

Exercise Studies in Research Journals

Name two research journals specializing in exercise physiology.

1.

2.

First Textbook in Exercise Physiology

Give two criteria for establishing a textbook as "first" in a field?

1.

2.

CONTRIBUTIONS OF THE HARVARD FATIGUE LABORATORY (1927-1946)

Name the first director of the Harvard Fatigue Laboratory.

Why was the Harvard Fatigue Laboratory important in the development of exercise physiology?

Other Early Exercise Physiology Research Laboratories

Name an "exercise physiology" department in the United States before 1935.

THE NORDIC CONNECTION (DENMARK, SWEDEN, NORWAY AND FINLAND)

Which of the Nordic countries first required physical training in the school curriculum?

Danish Influence

Name one famous Danish exercise physiologist.

Swedish Influence

Name one famous Swedish exercise physiologist.

Norwegian and Finnish Influence

Name one famous Norwegian exercise physiologist.

OTHER CONTRIBUTORS TO EXERCISE PHYSIOLOGY KNOWLEDGE

Name the most famous "physical fitness" researcher from the U.S.

CONTEMPORARY DEVELOPMENTS

Exercise Physiology and the World Wide Web

List two discussion groups of exercise physiology on the WWW.

1.

2.

Professional Exercise Physiology Organizations

Name two professional organizations that service exercise physiology/exercise science professionals.

1.

2.

COMMON LINK

Name two professors in your department that serve as mentor's for graduate students.

1.

2.

THE SCIENTIFIC METHOD AND EXERCISE PHYSIOLOGY

General Goals of Science

List two opposing views for the goals of science.

1.

2.

List three aims of science.

1.

2.

3.

Hierarchy in Science

List three levels of conceptualization of science.

1.

2.

3.

Fact Finding

Describe how to establish facts.

Interpreting Facts

What constitutes a variable?

List three categories of continuous variables.

1.

2.

3.

How does a discrete variable differ from a continuous variable?

CASUAL AND CAUSAL RELATIONSHIPS

Comment about the following statement: "Correlation does not imply causation."

Independent and Dependent Variables

Describe differences between independent and dependent variables.

Establishing Causality Between Variables

List and briefly describe two types of studies scientists use to establish cause and effect relationships between independent and dependent variables.

1.

2.

Experimental Studies

Describe the basis of an experiment.

List one key feature of experimental research.

Field Studies

Describe the essence of field study research.

FACTORS AFFECTING RELATIONSHIPS AMONG VARIABLES

Experimental Testing Effects

Describe the function of a control group in experimental research?

Measurement Errors

Describe two categories of measurement error.

1.

2.

Within-Subject (Intra-Individual) Variability

Describe the major difference between within-subject variability and individual differences.

Experimenter Expectation Effect

Describe how to minimize the Hawthorne effect.

ESTABLISHING LAWS

What roles do "laws" play in science?

Why are some laws better than others?

DEVELOPING THEORIES

List three components of a theory.

1.

2.

3.

Hypothetical Constructs

Give an example of an hypothetical construct.

Association Among Constructs

Give an example of a more clearly defined construct for "physical fitness"?

Operational Definitions

What role do operational definitions play in the scientific process?

Operationally define the follow:

1. IQ:

2. Upper arm strength:

3. Obesity:

CERTAINTY OF SCIENCE

Explain the statement, "failure to reject an hypothesis."

PUBLISHING RESULTS OF EXPERIMENTS

Briefly explain the importance of "peer review" in science.

EMPIRICAL VS. THEORETICAL — BASIC VS. APPLIED RESEARCH

Describe differences between empirical and theoretical research.

Describe differences between basic and applied research.

Practice Quiz

MULTIPLE CHOICE

1. Exercise physiology:
    a. Cross-disciplinary science
    b. Discrete science
    c. Not a bone fide field of study, but rather a subdiscipline of physiology
    d. Only deals with human research

2. The first "sports medicine" physician was:
    a. Austin Flint
    b. Edward Hitchcock
    c. Galen
    d. Hippocrates
    e. None of the above

3. The first professor of "exercise science" in the U.S. was:
    a. Austin Flint
    b. George W. Fitz
    c. J.D. Hooker
    d. Thomas K. Cureton
    e. George Williams

4. The first formal "exercise science" major:
    a. Harvard University, 1891
    b. Harvard Fatigue Laboratory, 1920
    c. George Washington University, 1800
    d. Amherst College, 1894
    e. None of the above

5. ACSM:
    a. Major professional group dedicated to professional training of physical therapists
    b. American College of Sports Medicine
    c. American College of Sports Marketing
    d. Original name for American Alliance for Health, Physical Education, Recreation and Dance
    e. b and c

6. A.V. Hill:
    a. Founder of ACSM
    b. Nobel laureate physiologist
    c. Founded Harvard Fatigue Lab
    d. Cofounder of the world wide web
    e. b and c

7. Facts:
    a. Of little importance to the scientific process
    b. Exhibit no moral quality
    c. "Building blocks" of science
    d. Do not require general agreement or consensus
    e. b and c

8. To infer causality between facts:
    a. A change in the dependent variable follows a change in the independent variable
    b. A change in the independent variable follows a change in the independent variable
    c. A change in the independent and dependent variables follow from a hypothesis
    d. Only a strong relationship between two variables must exist

9. Hawthorne effect:
    a. Introducing a change in work place conditions produces noticeable improvements in performance
    b. Introducing a change in work place conditions produces noticeable decrements in performance
    c. Researchers ability to minimize within-subject variability
    d. Researchers ability to control the placebo effect

10. Laws:
    a. Provide an explanation of why variables behave the way they do
    b. Provide a general summary of the relationship among variables
    c. Offer abstract explanation of facts
    d. Generate from inductive reasoning
    e. b and d

T R U E / F A L S E

1. _____ Facts exhibit no moral quality.

2. _____ Independent variables result from dependent variables.

3. _____ Field studies mostly investigate events as they occur in normal living.

4. _____ Theories attempt to explain the fundamental nature of laws.

5. _____ Empiricists collect facts and make observations with little regard for theory building.

6. _____ The first American exercise physiologist was D.B. Dill.

7. _____ Explaining, understanding, predicting, and controlling represent goals of science.

8. _____ The variable "male" is a discrete variable.

9. _____ The U.S. has always had licensing laws for physicians.

10. _____ The Harvard Fatigue Laboratory was initially housed at Princeton University, but staffed mainly by Harvard faculty.

CROSSWORD PUZZLE

ACROSS

CHAPTER 1

2. usually represents the quality predicted; presumed result
5. research with no concern for immediate practical application of research findings
6. type of variation that includes inherent tendency for humans to exhibit a variable response from moment-to-moment and trial-to-trial
8. largest professional organization in the world for exercise physiology (including allied medical and health areas) (abbr)
10. first US medical school
11. in 1898 three articles on physical activity appeared in the first volume (abbr)
12. developed Swedish "medical gymnastics," which became part of Sweden's school curriculum and influenced USA schoold PE programs
16. existence of true biological differences among individuals

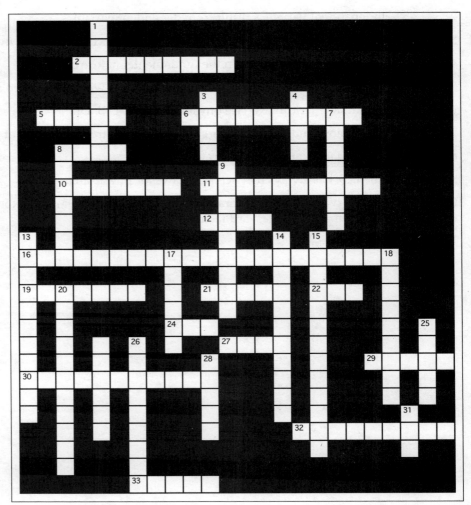

19. the basic source materials of historical research, includes testimony from reliable eye- and ear-witnesses to past events and direct examination of actual "objects" used in the past
21. observed phenomena through visual, auditory, and tactile sensory input
22. most famous laboratory in USA started in 1927 (abbr)
24. a statement describing the relationships among independent and dependent variables
27. official journal of American College of Sports Medicine (abbr)
29. perhaps the most well known and influential physician that ever lived; essays about exercise

and its effects might be considered the first formal "How To" manuals about such topics that remained influential for the next 15 centuries

30. research that emanates from and contribute to theory building
32. effect demonstrating that introducing any change can produce noticeable productivity increases
33. Danish Nobel laurate (discoved mechanism for capillary control of blood flow in resting and exercising muscle); pioneered studies of the relative contribution of fat and carbohydrate oxidation during exercise;

DOWN

1. attempts to explain the fundamental nature of laws; offer abstract explanations of laws and facts
3. developed the first formal exercise physiology laboratory and degree program at Harvard University
4   Federation Internationale de Medicine Sportive, comprised of the national sports medicine associations of over 100 countries (abbr)
7. research that incorporates scientific endeavors to solve specific problems
8. important American PE organization (abbr)
9. research with little regard for building theory
13. father of preventive medicine who contributed 87 treatises on medicine including several on health and hygiene
14. error that can be categorized into technological and recording error
15. error that includes inherent instability of measuring devices

17. British Nobel laurate; pioneered studies of oxygen uptake during exercise and recovery
18. information provided by a person who does not directly observe the event, object, or condition
20. usually becomes the predictor variable; presumed cause
23. Swidish "superstar" exercise physiologish; most famous graduate of the Swedish College of Physical Education
25. research that mostly investigate events as they occur in normal living
26. Professor of Hygiene and Physical Education at Amherst College; his professorship became the second such appointment in physical education in USA
28. 1836–1915 pioneer American physician-scientist, contributed significantly to the burgeoning literature in physiology
31. internet address (abbr)

Macronutrients and Food Energy

Define Key Terms and Concepts

1. 4.2 kcal·gm^{-1}

2. 5.65 kcal·gm^{-1}

3. 9.4 kcal·gm^{-1}

4. Alanine-glucose cycle

5. Apoprotein

6. Atwater factors

7. Bomb calorimeter

8. Bonking

9. Calorie

10. Carbohydrates

11. Cholesterol

12. Chylomicrons

13. Complete proteins

14. Complex carbohydrates

15. Deamination

16. Derived fats

17. Digestive efficiency

18. Direct calorimetry

19. Disaccharides

20. Essential amino acids

21. Fatty acids

22. Fiber

23. Glucagon

24. Glucogenesis

25. Gluconeogenesis

26. Glycerol

27. Glycogen

28. Glycogenolysis

29. Glycolipids

30. HDL

31. Heat of combustion

32. Hydrolysis

33. Hypoglycemia

34. Insulin

35. Kilogram calorie

36. Lactovegetarian

37. LDL

38. Lipase

39. Lipoproteins

40. Monosaccharides

41. Monounsaturated fatty acids

42. Neutral fats

43. Nonessential amino acids

44. Ovolactovegetarian

45. Peptide bonds

46. Phospholipids

47. Polysaccharides

48. Polyunsaturated fatty acids

49. Protein RDA

50. RDA

51. Reverse transfer of cholesterol

52. Saturated fatty acids

53. Starch

54. Transamination

55. Unsaturated fatty acids

56. Vegan

57. VLDL

Study Questions

CARBOHYDRATES

Monosaccharides

List three common monosaccharides and give a food source for each.

Monosaccharide   *Food Source*

1.

2.

3.

List three functions of glucose after absorbtion.

1.

2.

3.

Disaccharides

Give the components of the following three common dietary disaccharides.

Sucrose = _____ + _____

Maltose = _____ + _____

Lactose = _____ + _____

How much simple sugar is consumed in the typical American diet?

Polysaccharides

List two classifications of polysaccharides.

1.

2.

Plant Polysaccharides

List two forms of plant polysaccharides.

1.

2.

STARCH

What is starch?

FIBER

What is fiber?

Give the possible beneficial role of dietary fiber on gastrointestinal function and blood cholesterol.

Gastrointestinal function:

Blood Cholesterol:

How much fiber should you consume in your daily diet?

Animal Polysaccharides

What is the major animal polysaccharide?

Approximately how much animal polysaccharide can an 80-kg person store?

What is the major role of animal polysaccharides in the body?

What roles do insulin and glucagon play in blood sugar regulation?

Insulin:

Glucagon:

Diet Affects Glycogen Stores

Give the body's upper limit for glycogen storage.

Carbohydrate's Role in the Body

List four major functions of carbohydrate in the body.

1.

2.

3.

4.

Recommended Carbohydrate Intake

What is the typical amount of carbohydrate consumed as a percent of total calories ingested?

What is the recommended carbohydrate intake (g per day) for a physically active man and woman?

Man:

Woman:

Carbohydrate Utilization in Exercise

Intense Exercise

What is the primary fuel during intense exercise of short duration?

Moderate and Prolonged Exercise

Detail the utilization of carbohydrate during the initial, middle, and end stages of moderate but prolonged exercise.

Initial stage:

Middle stage:

End of exercise:

LIPIDS

Give a typical hydrogen-to-oxygen ratio for a common lipid.

Simple Lipids

What is another name for simple fats? Name the most common simple fat in the body.

Name:

Most common:

Saturated Fatty Acids

Define "saturated fatty acid."

List three food sources rich in saturated fatty acids.

1.

2.

3.

Unsaturated Fatty Acids

Define "unsaturated fatty acid."

List three food sources rich in unsaturated fatty acids.

1.

2.

3.

Fatty Acids in the Diet

What percent of the total caloric intake of the typical Americans diet consists of saturated fat?

Compound Lipids

List three groups of compound fats.

1.

2.

3.

High- and Low-Density Lipoprotein Cholesterol

List three types of lipoproteins?

1.

2.

3.

"Bad" Cholesterol/ "Good" Cholesterol

Identify the proper name for the "bad" and "good" form of cholesterol."

    Bad cholesterol:

    Good cholesterol:

Derived Lipids

What is the most widely known derived lipid?

How much endogenous cholesterol does the body normally produce?

List three major dietary sources of cholesterol?

1.

2.

3.

Lipids in Food

Plant and animal products typically contribute what percentage of the daily lipid intake?

    Plant:

    Animal:

Lipids' Role in the Body

List three major functions of lipids in humans.

1.

2.

3.

What aspect of the chemical structure of lipid causes it to contain more energy than an equal weight of carbohydrate?

List four fat-soluble vitamins.

1.

2.

3.

4.

Recommended Lipid Intake

As a percent of total calories, how much lipid should you consume in your diet?

How much cholesterol should you consume in your diet?

Fat Use in Exercise

Describe the uptake of fatty acids by active muscles during a prolonged bout of moderate exercise. (Hint: refer to Fig. 2.8 in your textbook.)

PROTEINS

How does the chemical structure of protein differ from lipid and carbohydrate?

Amino Acids

Draw the basic structure of an amino acid molecule. (Hint: refer to Fig. 2.9 in your textbook.)

The joining of two amino acids forms a _____.

The joining of three amino acids forms a _____.

Linkage of four or more amino acids forms a _____.

Essential and Nonessential Amino Acids

Explain the primary difference between an essential and nonessential amino acid.

Sources of Proteins

Dietary Sources

List three of the largest sources of dietary protein in the American diet.

1.

2.

3.

What does "biologic value" of a food refer to?

Synthesis in the Body

Synthesis of nonessential amino acids from the carbon , oxygen and hydrogen fragments of lipids and

carbohydrate's is called _____.

The process that removes an amine group from an amino acid molecule is called

_____.

Protein's Role in the Body

To what does the "anabolic" role of protein in body processes refer?

List three important functions of protein in the body.

1.

2.

3.

Vegetarian Approach to Sound Nutrition

List two potential nutritional problem areas for men and women on a vegetarian diet.

1.

2.

Why would a "strict" vegetarian consider switching to a lactovegetarian or ovolactovegetarin diet?

Recommended Dietary Allowance

Why should a person regularly consume the RDA for a particular nutrient?

What is the RDA for protein based on body weight?

  Per kg weight:

  Per pound of weight:

Compute the protein RDA for a person who weighs 122 lb.

Explain why some athletes may have a greater requirement for protein than non-athletes?

Protein Requirements for Physically Active People

Do athletes require more protein?

What are the protein recommendations (per kg body weight) for athletes in heavy training?

Alanine-Glucose Cycle

Diagram the alanine-glucose cycle. (Hint: refer to Fig. 2.12 in your textbook.)

Energy derived from the alanine-glucose cycle may supply what percentage of the total exercise energy requirement, and what percentage of the liver's glucose output?

Exercise energy requirement:

Liver's glucose output:

ENERGY CONTENT OF FOOD

Calorie —A Measurement Unit of Food Energy

Describe the difference between a calorie and a kilocalorie.

Gross Energy Value of Foods

Describe the instrument in direct calorimetry to measure a food's energy content?

Heat of Combustion

Give the heats of combustion for the three macronutrients:

Carbohydrate:

Lipid:

Protein:

Why is the heat of combustion for protein less in the body than in a bomb calorimeter?

Net Energy Value of Foods

Compare the heat of combustion of carbohydrates and lipids in the body and determined by bomb calorimetry.

In the body: *CHO*   *LIPID*

Bomb calorimetry:

Digestive Efficiency

List one effect that dietary fiber has on the energy availability of ingested foods?

Considering digestive efficiency, the net kcal value for the three macronutrients equals:

Carbohydrate:

Lipid:

Protein:

Energy Value of a Meal

Calculate the caloric content of 100 grams of a food containing 5% protein, 14% lipid, and 20% carbohydrate. (Hint: Use the Atwater general factors.)

Calories Equal Calories

From an energy standpoint, explain why 100 calories from a piece of cake is no more fattening that 100 kcal from celery?

Practice Quiz

MULTIPLE CHOICE

1. Which of the following is not a pathway for glucose after it is absorbed by the small intestine?
   a. Converted to fats for energy storage
   b. Stored as glycogen in the muscles and liver
   c. Used by the tissues to preserve protein
   d. Used directly by the cell for energy
   e. All of the above

2. Disaccharides:
   a. Double sugars
   b. Sucrose, fructose, and maltose
   c. Form from combination of starch and fiber
   d. All contain fructose
   e. a and b

3. Reconverting glycogen to glucose is the process of:
   a. Deamination
   b. Gluconeogenesis
   c. Glycogenolysis
   d. Transamination
   e. None of the above

4. The main function of carbohydrate:
   a. Spare protein
   b. Provide insulation and protection for vital organs
   c. Serve as an energy fuel for the body
   d. Prevent hypoglycemia
   e. None of the above

5. Cholesterol:
   a. Is a derived lipid
   b. Is classified into three types: HDL, MDL and GDL
   c. Is found in foods of plant and animal origin
   d. Is necessary for the synthesis of vitamins A and B
   e. All of the above

6. A kilogram-calorie:
   a. Amount of heat liberated from the body
   b. Amount of heat necessary to heat a nutrient 1°C
   c. Amount of heat necessary to raise the temperature of 1 kg of water 1°C
   d. All of the above
   e. None of the above

7. Which statement is true about nutrient digestive efficiency?
   a. 97% for carbohydrates, 95% for lipids, and 92% for proteins
   b. 97% for lipids, 95% for carbohydrates, and 92% for proteins
   c. 97% for proteins, 95% for carbohydrates, and 92% for lipids
   d. 97% for lipids, 95% for proteins, and 92% for carbohydrates
   e. None of the above

8. Which statement is true about a bomb calorimeter?
   a. Produces Atwater Factors
   b. Used in direct calorimetry
   c. Used in indirect calorimetry
   d. Determines the digestive efficiency of food

9. What is the average net energy value for carbohydrate, lipid, and protein, respectively?
   a. 4, 4, 9
   b. 9, 9, 4
   c. 4, 9, 4
   d. 9, 4, 9
   e. None of the above

10. Which food has the greatest energy content?
    a. 250 kcal of chocolate
    b. 350 kcal of milk
    c. 300 kcal of ice cream
    d. 275 kcal of pure oil
    e. None of the above

T R U E / F A L S E

1. _____ Lipid intake in a prudent diet should not exceed 30% of the total energy content of the diet.

2. _____ Adequate carbohydrate intake will have a sparing effect on the body's lean tissue.

3. _____ Most vegetable oils are relatively high in unsaturated fatty acids.

4. _____ Cholesterol cannot be produced in the body.

5. _____ Most common proteins contain approximately 16% nitrogen.

6. _____ A calorie is a measure used to express the heat or energy value of food.

7. _____ A"prudent diet" should contain about 60% of its calories in the form of protein.

8. _____ Fruits are generally digested more rapidly than eggs.

9. _____ Atwater general factors refer to the average net energy value of the macronutrients.

10. _____ Protein breakdown during exercise is less in lean compared with obese individuals.

CROSSWORD PUZZLE

CHAPTER 2

ACROSS

1. Breakdown of glycogen
3. dextrose or blood sugar, consists of 6 carbon, 12 hydrogen, and 6 oxygen atoms C6H12O6
5. Monosaccharide not found freely in nature
6. Glucose + glucose
7. Form of chemical bonding linking amino acids
9. Chemical breakdown by addition of water
12. Glycogen formation from glucose
14. Low blood sugar
15. high blood sugar
17. The sweetest monosaccharide
20. contain three or more simple sugars
22. Nondigestible plant component
23. Derived lipid synthesized by liver
26. Vegetarian who eats eggs and drinks milk
28. facilitates the muscle cell's uptake of glucose excess
33. Gluconeogenic amino acid
34. Transport form for blood lipids
37. insulin antagonist hormone
39. Carbon containing compounds
40. Fructose + glucose
41. Net energy value for macronutrients named after this chemist
42. Good cholesterol (abbr.)
43. Macronutrients insoluble in H20
44. true vegetarians
45. Bad cholesterol (abbr.)
46. originates from plant sources; made up of atoms of carbon,

DOWN

2. Milk sugar
4. breakdown of substance
5. Sugar component of triglycerides
6. basic unit of carbohydrates
8. Glucose synthesis from noncarbohydrate sources
10. Less than 10 g fat, 4.5 g saturated fat, and 95 mg cholesterol per serving, and per 100 g of meat, poultry, and seafood

11. combinations of two monosaccharides
13. form after emulsified lipid droplets leave the intestine and enter the lymphatic vasculature
16. Removing nitrogen from amino acids
18. A liberal standard for nutrient intake
19. 50% or less fat than comparison food (e.g., "50% less fat than our regular cookies"
21. Lipoprotein with largest amount of lipid (abbr.)
24. A _____ fatty acid contains one double bond along the main carbon chain
25. Less than 0.5 g fat per serving (no added fat or oil)
27. Glycerol + 3 free fatty acids
29. Fatty acids + phosphorous containing compound
30. Protein building block
31. _____ calorimeter measures energy value of food
32. Animal polysaccharide
35. Less than 5 g fat, 2 saturated fat and 95 mg cholesterol per serving and per 100 g of meat, poultry, and seafood
36. No double bonds in this type of fatty acid
38. enzymes that breaks down lipid
40. A complex plant carbohydrate

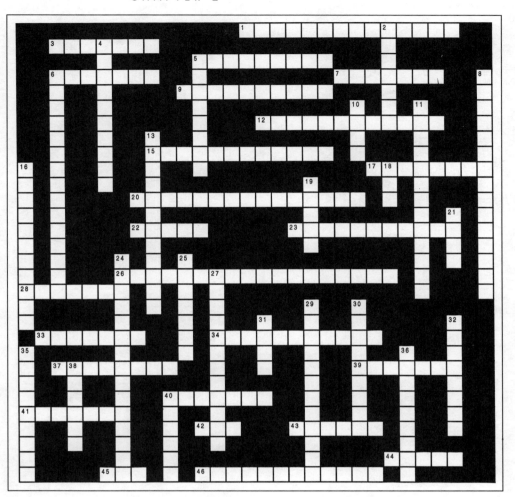

Micronutrients and Water

1. Aldosterone

2. Antioxidants

3. B-complex vitamin

4. Bone remodeling

5. Calcatonin

6. Carotenes

7. Clinical anemia

8. Coenzymes

9. Cytochromes

10. Electrolytes

11. Exertional heat stroke

12. Extracellular fluid

13. Fat-soluble vitamins

14. Free radicals

15. Heat cramps

16. Heat exhaustion

17. Heat illness

18. Heme iron

19. Hemoglobin

20. Hypervitaminosis A

21. Hyponatremia

22. Insensible perspiration

23. Intracellular fluid

24. Iron deficiency anemia

25. Lipid peroxidation

26. Major minerals

27. Megavitamins

28. Metabolic Water

29. Micronutrients

30. Minerals

31. Myoglobin

32. Non-heme iron

33. Osteoporosis

34. Oxidative stress

35. Provitamin

36. Relative humidity

37. Secondary amenorrhea

38. Sodium-induced hypertension

39. Sports anemia

40. β-carotene

41. Trace minerals

42. Transferrin

43. Vitamins

44. Water-soluble vitamins

Study Questions

MICRONUTRIENTS

Vitamins

What was the first vitamin discovered?

How does the body acquire vitamins?

Classification of Vitamins

How many vitamins have been isolated and RDA levels established?

What differences exist between a vitamin obtained naturally from food and a vitamin produced synthetically?

FAT-SOLUBLE VITAMINS

List four fat-soluble vitamins.

1.

2.

3.

4.

WATER-SOLUBLE VITAMINS

List nine water-soluble vitamins.

1.   6.

2.   7.

3.   8.

4.   9.

5.

Potential Toxicity of Vitamins

Discuss whether water-soluble or fat-soluble vitamins are potentially harmful when taken in excess.

Vitamins' Role in the Body

List two major functions of the fat-soluble vitamins?

1.

2.

Do athletes require more vitamins than sedentary people? Explain.

Antioxidant Role of Specific Vitamins

Which vitamins protect against oxidative stress?

List three major antioxidant enzymes.

1.

2.

3.

Vitamins and Exercise Performance

Do physically active individuals require increased amounts of antioxidant substances? Explain.

Discuss whether vitamins provide "quick energy" for exercise.

Vitamin Supplements: The Competitive Edge?

Do vitamin supplements enhance physical performance? Explain.

Megavitamins

Discuss whether a rationale exists for consuming megadoses of vitamins to enhance exercise performance.

In what ways can excessive vitamin intake harm the body?

Vitamins Behave as Chemicals

Give an example of the effects of excesses of a water-soluble and a fat-soluble vitamin on bodily function.

Water-soluble vitamin:

Fat-soluble vitamin:

MINERALS

The Nature of Minerals

What percent of the body's mass consists of minerals?

The body is composed of how many minerals?

List four minerals and their function in the body.

Mineral   *Function*

1.

2.

3.

4.

What is the exception to the phrase: "There is generally adequate mineral intake"?

Kinds, Sources, and Functions of Minerals

List two major functions of minerals in the the body.

1.

2.

Why is iodine important to bodily function?

Minerals and Physical Activity

Calcium

List five functions of calcium in the body.

1.   4.

2.   5.

3.

OSTEOPOROSIS, CALCIUM, ESTROGEN, AND EXERCISE

What happens to bones without sufficient calcium intake?

How can osteoporosis be prevented?

List two factors that contribute to "porous" bone.

1.

2.

DIETARY CALCIUM CRUCIAL

Give the calcium RDA for the following individuals:

Adolescents and young adults:

Adults over age 24:

Women after menopause:

Exercise Helps

Compared with nonexercisers, what part of the bone increases in people who exercise regularly regardless of age?

List three principles for planning exercise to promote bone health.

1.

2.

3.

Female Athlete Triad: An Unexpected Problem for Women Who Train Intensely

Explain how too much exercise can be counterproductive in preventing osteoporosis.

Describe the three parts of the female "triad."

1.

2.

3.

Sodium, Potassium, and Chlorine

What role does sodium and chlorine play in the body's fluid regulation?

List two functions of sodium and potassium.

Sodium   **Potassium**

1.

2.

SODIUM: HOW MUCH IS ENOUGH?

List the recommended daily intake of sodium.

List two possible negative effects of excess sodium intake.

1.

2.

Iron

List four functions of iron in the body.

1.   3.

2.   4.

IRON STORES

List two dietary sources of iron.

1.

2.

Describe "iron deficiency anemia."

OF CONCERN TO VEGETARIANS

Why are vegetarians at greater risk for developing iron insufficiency?

FEMALES: A POPULATION AT RISK

Discuss the reasoning behind considering females an "at risk" population for iron insufficiency.

EXERCISE-INDUCED ANEMIA: FACT OR FICTION?

Describe the condition of "sports anemia."

List three factors related to "sports anemia."

1.

2.

3.

ATHLETES AND IRON SUPPLEMENTS

Do athletes (particularly females) who consume a balanced diet with adequate calories require added iron supplements?

Minerals and Exercise Performance

Can mineral supplements consumed above recommended levels on a continual basis enhance exercise performance or training responsiveness?

Defense Against Mineral Loss

Under what conditions would you recommend "athletic drinks" to supplement minerals lost through exercising?

WATER

Water constitutes what percent of the weight of muscle tissue and of body fat weight?

Weight of muscle:

Weight of body fat:

Functions of Body Water

List five functions of water in the body.

1.   4.

2.   5.

3.

Water Balance: Intake Versus Output

Water Intake

List and quantify three main sources of water intake.

Source	*Amount*
1.	
2.	
3.	

Water Output

List and quantify four primary ways the body loses water.

Source	*Amount*
1.	
2.	
3.	
4.	

Water Requirement in Exercise

List three factors influencing water loss through sweating during exercise.

1.

2.

3.

Heat Disorders

Heat Cramps

Describe symptoms of heat cramps and the conditions that precipitate it.

Heat Exhaustion

Describe symptoms of heat exhaustion and the conditions that precipitate it.

Exertional Heat Stroke

Describe symptoms of exertional heat stroke and the conditions that precipitate it.

Practical Recommendations for Fluid Replacement in Exercise

List three reasons for ingesting extra water before exercising in the heat.

1.

2.

3.

Gastric Emptying

List three factors that influence gastric emptying.

1.

2.

3.

Adequacy of Rehydration

List two practical recommendations to ensure adequate body water levels in athletes.

1.

2.

Water Intoxication

Describe the condition of hyponatremia.

List two factors predispose an athlete to develop hyponatremia?

1.

2.

Practice Quiz

MULTIPLE CHOICE

1. Vitamins are important components in physiologic processes related to:
   a. Fluid osmolality
   b. Metabolic regulation
   c. Temperature regulation
   d. Insulation
   e. c and d

2. Minerals:
   a. Function in anabolic and catabolic metabolic process
   b. Primarily come from animal dietary sources
   c. Are scarce in the plant kingdom
   d. Alone can catalyze metabolic reactions
   e. a and d

3. These three main sources supply the water needs of the body:
   a. Liquids, fruits, vegetables
   b. Foods, fluids, metabolism
   c. Fruits, liquids, solids
   d. Lipids, proteins, carbohydrates
   e. None of the above

4. In extreme hot weather, the fluid needs of an active person increase by _____ x:
   a. 2
   b. 3
   c. 6
   d. 8
   e. 10

5. Excess vitamin B_6 intake:
   a. Can lead to liver disease and nerve damage
   b. Has no known side effects
   c. Accelerates osteoporosis
   d. Stunts growth if accompanied by excessive vitamin B12 intake
   e. Can cause muscle atrophy

6. Provitamins:
   a. Facilitate cholesterol removal from arterial walls
   b. Activate certain vitamins
   c. Are composed primarily of lipid soluble products
   d. Are precursor substances found only in animals
   e. Represent vitamin supplements advertised by athletes

7. The two most plentiful minerals in the body:
   a. Calcium, phosphorus
   b. Iron, selenium
   c. Calcium, magnesium
   d. Chlorine, potassium
   e. Calcium, iron

8. Women on vegetarian-type diets:
   a. Have a higher risk of developing iron insufficiency
   b. Have a lower risk of developing iron insufficiency
   c. Need to decrease their intake of vitamin C-rich foods
   d. Need to increase their intake of fat-soluble vitamins to increase heme iron availability
   e. c and d

9. Megadoses of vitamin C:
   a. Raise serum uric acid levels and precipitate gout in people predisposed to this disease
   b. Are not dangerous because any excess is voided in urine
   c. Can cause hypermetabolism during rest
   d. Enhances aerobic endurance performance
   e. None of the above

10. Greatest loss of daily water occurs during exercise:
    a. Urine
    b. Feces
    c. Skin
    d. Pulmonary ventilation
    e. Insensible perspiration

T R U E / F A L S E

1. _____ A, D, E, and K represent the fat-soluble vitamins.

2. _____ B-complex vitamin supplementation above the RDA improves aerobic exercise performance.

3. _____ Mineral intake is usually adequate when consuming a well-balanced diet.

4. _____ Water makes up between 40 to 60% of the total body mass.

5. _____ The primary aim of fluid replacement is to maintain plasma volume.

6. _____ All vitamins except vitamin E can be synthesized in the body.

7. _____ Sports anemia occurs when hemoglobin in men decreases to about 14 g of hemoglobin per 100 mL of blood.

8. _____ Extracellular water volume exceeds the intracellular volume of water.

9. _____ Na, K, and Cl are collectively called erythrocytes.

10. _____ Osteoporosis begins early in life due mainly to inadequate calcium intake.

CROSSWORD PUZZLE

CHAPTER 3

ACROSS

4. Pigmented respiratory protein in erythrocytes
5. hyperpyrexia (rectal temp >41C, lack of sweating, disorientation, twitching, seizures, coma due to extreme hyperthermia
7. Absence of menses
8. Vitamin precursors in yellow vegetables are Beta _____
11. plasma protein that transports iron from ingested food and damaged red blood cells for delivery to tissues in need
13. best known of the pigmented compounds, or carotenoids, that give color to yellow, orange, and green, leafy vegetables and fruits
14. minerals required in amounts greater than 100 mg a day
15. reduced hemoglobin levels approaching clinical anemia (12 g per 100 mL of blood for women and 14 g per 100 mL blood for men
18. tightening cramps, involuntary spasms of active muscles; low serum Na+ due to intense prolonged exercise in the heat
19. Dietary component required in small quantities
23. clinical condition related to low hemoglobin concentration
29. Salt regulating hormone of adrenal cortex
30. organic substances that neither supply energy nor contribute to body mass
31. hypohydration, flushed skin, reduced sweating in extreme dehydration syncope due to cumulative negative water balance
32. hormone that minimizes sodium and water loss through the kidneys and in sweat

DOWN

1. vitamin doses of at least tenfold and up to 1000 times the RDA
2. specialized substances that facilitate cellular energy transfer
3. Oxygen carrying group of hemoglobin
6. Outside cell
9. vitamins in food remaining in an inactive or precursor form
10. Minerals required in amounts less than 100 mg/day
12. electrically charged particles
16. iron from plant sources
17. menstrual cycle ceases
19. similar to hemoglobin that aids in oxygen storage and transport within muscle cells
20. process of gaining water from a hypohydrated state towards duhydration
21. small quantities of vitamins and minerals that play highly specific roles to facilitate energy transfer
22. water intoxication when serum sodium concentration falls below 136 mEq·L-1
23. _____ perspiration
24. disease typified by reduction in bone mass
25. Inside cell
26. Approximately 4% of the body's mass consistint of some 22 mostly metallic elements
27. low hemoglobin
28. process of losing water

Fundamentals of Human Energy Transfer

1. Acetyl-CoA

2. Acid

3. Acidosis

4. Adipocytes

5. ADP

6. Aerobic glycolysis

7. Aerobic metabolism

8. Alkalosis

9. AMP

10. Anaerobic glycolysis

11. Anaerobic reaction

17. Citrate

12. ATP

18. Coenzyme A

13. ATPase

19. Cori cycle

14. Base

20. Coupled reactions

15. Buffers

21. PCr

16. Chemical buffers

22. Creatine kinase

23. Cyclic AMP

24. Cytochrome oxidase

25. Cytochromes

26. Deamination

27. Electron transport

28. FAD

29. Free fatty acids

30. Glucose 6-phosphate

31. Glycerol

32. Glycogen phosphorylase

33. Glycogen synthetase

34. Glycogenolysis

35. Glycolysis

36. High-energy phosphate

37. Hydrolysis

38. Ketone bodies

39. Krebs Cycle

40. Lactic acid

41. Lactic dehydrogenase

42. Lipase

43. Lipolysis

44. Lipoprotein lipase

45. Mitochondria

46. NAD$^+$

47. Oxaloacetate

48. Oxidative phosphorylation

49. P÷O ratio

50. pH

51. Phosphatase

52. Phosphate bond energy

53. Phosphofructokinase

54. Phosphorylation

55. Renal buffer

56. Respiratory chain

57. β-oxidation

58. Substrate-level phosphorylation

59. The metabolic mill

61. Ventilatory buffer

60. Transamination

PHOSPHATE BOND ENERGY

Adenosine Triphosphate: The Energy Currency

List two major energy-transforming activities in a cell.

1.

2.

Complete the reaction:

$ATP + H_2O \longrightarrow$

Describe the chemical make-up of ATP.

Phosphocreatine: The Energy Reservoir

What main function does PCr play in energy metabolism?

ATP: A Limited Currency

How much ATP can the body store?

Complete the following two equations.

$$ATP + H_2O \xrightarrow{\text{ATPase}}$$

$$PCr + H_2O \xrightarrow{\text{Creatine kinase}}$$

Intramuscular High-Energy Phosphates

How long can the intramuscular energy-rich phosphates ATP and PCr sustain all-out exercise?

ENERGY SOURCE IMPORTANT

Chemical Bonds Transfer Energy

Name the process for transferring energy in the form of phosphate bonds.

CELLULAR OXIDATION

"For every reaction involving cellular oxidation, there is a reaction involving _____

_____."

Electron Transport

List the two electron acceptors in food oxidation during energy metabolism.

1.

2.

Oxidative Phosphorylation

Describe oxidative phosphorylation in 30 words or less.

Complete the following chemical equation:

$$NADH + H^+ + 3ADP + 3P_i + 1/2\,O_2 \rightarrow$$

Efficiency of Electron Transport and Oxidative Phosphorylation

How many kcal are required to form one mole of ATP?

How many moles of ATP are resynthesized for each mole of NADH oxidized?

Role of Oxygen in Energy Metabolism

List three prerequisites for continual resynthesis of ATP during aerobic metabolism.

1.

2.

3.

"The main role of oxygen in energy metabolism is to _____

_____ "

ENERGY RELEASE FROM FOOD

Macronutrient breakdown serves what crucial purpose?

ENERGY RELEASE FROM CARBOHYDRATE

Write the equation for the complete breakdown (hydrolysis) of one mole of glucose.

Give the efficiency of carbohydrate breakdown for conserving phosphate-bond energy?

The net ATP yield from the complete breakdown of one mole of glucose equals _____.

Anaerobic Versus Aerobic

The two stages of carbohydrate breakdown are called _____ and

_____ .

Anaerobic Energy from Glucose: Glycolysis (Glucose Splitting)

How many chemical reactions are there in the breakdown of glucose in glycolysis?

Name the process of glycolysis when it begins with stored glycogen _____.

The rate-limiting enzyme in the anaerobic breakdown of glucose is _____.

List two enzymes in glycolysis.

1.

2.

Glycolysis occurs in what part of the cell?

Substrate-Level Phosphorylation

Substrate-level phosphorylation in glycolysis resynthesizes how many ATPs?

  Gross gain _____

  Net gain _____

Anaerobic glycolysis generates what percentage of the total ATP from glucose breakdown?

Hydrogen Release in Glycolysis

How many ATP form when cytoplasmic NADH in glycolysis oxidizes in the respiratory chain?

Lactate Formation

Name the end product of "aerobic glycolysis"?

Identify these two substances.

$C_3H_4O_3$:

$C_3H_6O_3$:

Write the chemical equations to illustrate the formation of lactic acid during exercise. (Hint: refer to Fig. 4.10 in your textbook)

List two fates for lactic acid in the body?

1.

2.

Aerobic Energy From Glucose: Krebs Cycle

Give the most important function of the Krebs cycle.

The Krebs cycle requires how many chemical steps?

How many ATP form directly from substrate phosphorylation in the Krebs cycle?

How many CO_2 and H_2 form for each molecule of acetyl-CoA that enters the Krebs cycle?

CO_2 _____

H_2 _____

Net Energy Transfer From Glucose Catabolism

Fill-in the following table to indicate the net ATP yield from the complete breakdown of glucose. (Hint: refer to Fig. 4.15 in your textbook.)

Source	Reaction	Net ATP
Substrate phosphorylation		
$2H_2$ (4 H)		
$2H_2$ (4 H)		
Substrate phosphorylation		
$8H_2$ (16H)		
	Total ATP _____	

ENERGY RELEASE FROM FAT

What enzyme catalyzes fat breakdown?

Complete the following equation:

Triglyceride + $3H_2O \rightarrow$

Adipocytes: Site of Fat Storage and Mobilization

List three hormones that augment FFA mobilization into the blood.

1.

2.

3.

What intracellular substance catalyzes lipase activation?

Breakdown of Glycerol and Fatty Acids

Which substance undergoes beta oxidation?

True or False: Fatty acid breakdown associates directly with oxygen uptake.

Glycerol

How many molecules of ATP synthesize when one glycerol molecule breaks down?

What is glycerol's important function in carbohydrate balance?

Fatty Acids

Of what importance is oxygen in fatty acid catabolism?

GLUCOSE NOT RETRIEVABLE FROM FATTY ACIDS

Why can't fatty acids provide energy for tissues that utilize glucose almost exclusively for fuel (e.g., brain and nerve tissue)?

Total Energy Transfer From Fat Catabolism

What is the efficiency of energy conservation for fatty acid oxidation?

How many molecules of ATP become synthesized in the complete combustion of a neutral fat molecule?

Fat Catabolism in Exercise

Intracellular and extracellular fat supply approximately how much of the energy (% range) for physical activity?

List three fat-burning adaptations within skeletal muscle with aerobic training.

1.

2.

3.

ENERGY RELEASE FROM PROTEIN

What two processes remove nitrogen from amino acids?

1.

2.

After nitrogen removal from an amino acid, what happens to the remaining "carbon skeleton" in energy metabolism?

THE METABOLIC MILL

Sketch the metabolic mill showing the primary interconversions among the nutrients. (Hint: refer to Fig. 4.20 in your textbook.)

FATS BURN IN A CARBOHYDRATE FLAME

Explain the quote, "fats burn in a carbohydrate flame."

Energy Releases More Slowly From Fat

Why would depleted muscle glycogen probably cause a substantial decrease in power output during prolonged exercise?

Excess Macronutrients Convert to Fat

Can protein consumed in excess of the body's energy requirement end up as stored fat? Explain.

REGULATION OF ENERGY METABOLISM

What exerts the greatest effect on the rate-limiting enzymes that control catabolism of the macronutrients?

List two rate limiting substances that control energy metabolism.

1.

2.

ACID-BASE REGULATION AND PH

Acid

Give three characteristics of an acid.

1.

2.

3.

Base

Give three characteristics of a base.

1.

2.

3.

pH

List two differences between an acid and a base.

1.

2.

List four substances and their corresponding pH values.

Substance	*pH*
1.	
2.	
3.	
4.	

Enzymes and pH

Describe the effect of pH on enzyme function.

Buffers

List three buffer mechanisms that regulate the body's acid-base balance.

1.

2.

3.

Chemical Buffers

Chemical buffers consist of a _____ acid and a _____ or _____ of that acid.

_____ represents another example of a chemical buffer.

Ventilatory Buffer

At any given work load, reducing pulmonary ventilation causes what change in the body's acid-base balance?

Renal Buffer

The renal buffer controls acidity through regulation of what substances?

BUFFERING AND STRENUOUS EXERCISE

Draw the relationship between lactic acid and blood pH during increasing intensities of short duration exercise up to maximum (Hint: Refer to Fig. 4.23 in your textbook).

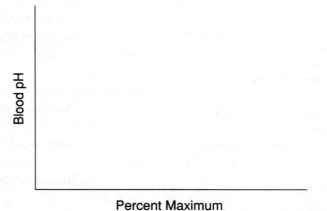

Practice Quiz

MULTIPLE CHOICE

1. How many kcals of free energy are liberated per mole of ATP hydrolyzed to ADP?
   - a. 6.3
   - b. 7.3
   - c. 8.3
   - d. 9.3
   - e. None of the above

2. The energy released from the breakdown of the intramuscular high-energy phosphates can sustain all-out exercise for about:
   - a. 5 to 8 s
   - b. 30 s
   - c. 60 to 90 s
   - d. 3 to 5 min
   - e. Longer than 5 min

3. Which is false about oxidative phosphorylation?
   a. ATP becomes synthesized during the transfer of electrons from $NADH_2$ and $FADH_2$ to oxygen
   b. About 50% of ATP synthesis occurs in the respiratory chain by oxidative reactions coupled with phosphorylation
   c. Three ATP form for each $NADH_2$ oxidized in the respiratory chain
   d. Two ATP form for each $FADH_2$ oxidized in the respiratory chain
   e. b and c

4. Which of the following is not a prerequisite for continual ATP resynthesis?
   a. PCr must be depleted in the cell
   b. $NADH_2$ or $FADH_2$ must be available
   c. Oxygen must be present in the tissues
   d. Cellular enzymes must be present in sufficient concentration
   e. a and c

5. Which is true about carbohydrates?
   a. Only nutrient whose stored energy generates ATP aerobically
   b. Supplies about one-half of the body's energy requirements during heavy exercise
   c. Functions as a metabolic primer for protein breakdown
   d. Continual carbohydrate breakdown required for fat to be used effectively for energy
   e. a and d

6. Net ATP yield from the complete breakdown of glucose in skeletal muscle:
   a. 28
   b. 32
   c. 36
   d. 38
   e. None of the above

7. Which is false concerning lactic acid?
   a. Formed during anaerobic glycolysis
   b. Formed by hydrogen's combining temporarily with pyruvate
   c. Buffered to lactate in the blood and carried away from the site of energy metabolism
   d. Valuable source of chemical energy that accumulates during heavy exercise
   e. By-product of excessive protein metabolism

8. Body's most plentiful source of potential energy:
   a. Glucose
   b. Proteins
   c. Fatty acids
   d. Minerals
   e. Glucose + glycogen

9. Which is true about glycerol?
   a. Accepted into glycolysis during its catabolism
   b. Enters the Krebs cycle directly
   c. Precursor of fatty acids
   d. Removes lactic acid from active muscle
   e. a and d

10. Nitrogen removal from amino acids:
    a. Deamination
    b. Translocation
    c. Gluconeogenesis
    d. Glycogenolysis
    e. The Cori cycle

T R U E / F A L S E

1. _____ Splitting an ATP molecule requires oxygen.

2. _____ Cells store PCr in larger quantities than ATP.

3. _____ NAD^+ and FAD accept hydrogens.

4. _____ The relative efficiency for harnessing chemical energy of electron transport-oxidative phosphorylation is approximately 90%.

5. _____ The activity level of glycogen synthetase places a limit on the rate of glycolysis during all-out exercise.

6. _____ Lactic acid accumulates when NADH oxidation keeps pace with its rate of formation.

7. _____ The Krebs cycle degrades acetyl-CoA substrate to carbon dioxide and nitrogen atoms within the mitochondria.

8. _____ As blood flow increases with exercise, adipose tissue releases more free fatty acids for delivery to active muscle.

9. _____ Fatty acids breakdown associates directly with oxygen uptake.

10. _____ Fats burn in the "flame" of carbohydrates.

CROSSWORD PUZZLE

ACROSS

4. fatty acids to acetyl compounds
5. addition of phosphate to an organic compound, such as glucose to produce glucose 6-phosphate
7. enzyme that catalyzes ATP catabolism
8. compound that regulate rate of metabolism
10. term relating to base
11. name for glucose to lactate reactions
12. binds with [H+]
13. continuous activities of 3 min and longer engage mainly this system energy transfer
14. removing nitrogen from amino acids
16. fat cell
19. marker enzyme for glycolytic capacity
23. compounds that minimize change in H+
25. electron transport _____
27. hydrogenaccepting coenzyme that contains niacin
28. product of incomplete lipid breakdown
30. involved in the transfer of acyl groups in energy metabolism
31. fat storing enzyme
33. intramitochondrial iron protein compound involved in hydrogen ion transfer
34. condition of abnormally high [H+]
35. breakdown of glycogen

DOWN

1. breakdown of lipid
2. "powerhouses" of the cell
3. process of removing nitrogen from an amino acid passing it to other compounds
5. phosphocreatine (abbr)

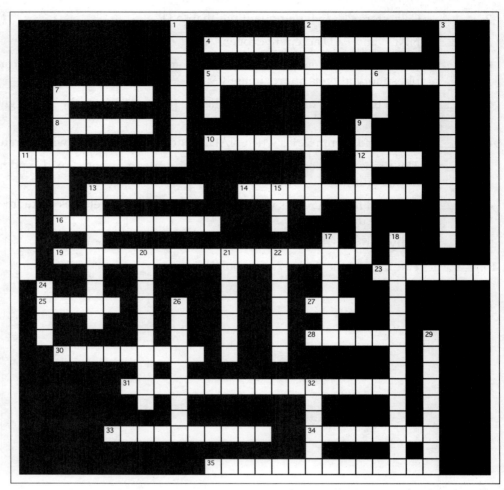

6. contains two high energy phosphate bonds
7. pyruvic acid + coenzyme A
9. chemical compound acted on by an enzyme
11. component of triglyceride molecule
13. protein building block
15. powers all biologic work
17. buffered form of lactic acid
18. glucose synthesis from noncarbohydrate sources
20. chemical breakdown by addition of water
21. _____ reactions occur in pairs; the breakdown of one compound provides energy for building another compound
22. abnormally high levels of ketone bodies
24. liberates H+ in solution
26. _____ capacity equals total energy generated by the immediate and short term energy system
29. condition of abnormally high pH
32. enzyme that catalyzes triglyceride breakdown

Human Energy Transfer During Exercise

Define Key Terms and Concepts

1. A.V. Hill

2. Active recovery

3. Alactacid oxygen debt

4. Blood lactate threshold

5. Creatine Kinase

6. Criteria for $\dot{V}O_{2max}$

7. EPOC

8. Ergometer

9. E-to-R method

10. Fast-twitch fibers

11. Immediate energy system

12. Interval training

13. Lactacid oxygen debt

14. Lactate dehydrogenase

15. Lactate shuttling

16. Lactate-using organs

17. Long term energy system

18. Maximal oxygen uptake

19. Oxygen debt

20. Oxygen deficit

21. Passive recovery

22. Short term energy system

23. Slow twitch fibers

25. Stored energy credits

24. Steady-rate

26. Stored phosphagens

Study Questions

IMMEDIATE ENERGY: THE ATP-CP SYSTEM

Indicate the quantity of ATP and PCr stored within the body's muscles.

ATP:

PCr:

Indicate the duration of a brisk walk, a slow run, and an all-out sprint that the intramuscular high energy phosphates powers.

Brisk walk:

Slow run:

All-out sprint:

SHORT-TERM ENERGY: THE LACTIC ACID SYSTEM

List three activities powered primarily by the lactic acid energy system.

1.

2.

3.

Blood Lactate Accumulation

At what percentage of the maximum oxygen uptake does blood lactate generally begin to accumulate?

What reasons might explain why the blood lactate threshold occurs at a lower or higher percent of $\dot{V}O_{2max}$ for aerobically trained compared to untrained subjects?

Give a reasonable explanation why blood lactate accumulates during exercise.

Lactate-Producing Capacity

How much more blood lactate can a sprint/power trained athlete accumulate compared with an untrained counterpart?

Blood Lactate as an Energy Source

List two tissue that use lactate as an energy source.

1.

2.

Explain the liver's role in lactate "removal."

LONG-TERM ENERGY: THE AEROBIC SYSTEM

Oxygen Uptake During Exercise

Draw and label the curve for oxygen uptake during 10 minutes of moderate running exercise.

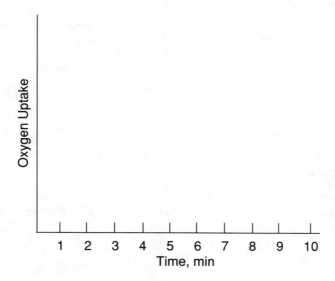

Indicate the area of oxygen deficit and the area of steady-rate. (Hint: Refer to Fig. 5.2 in your textbook.)

Express oxygen uptake in relation to body mass for an individual who weighs 85 kg and consumes 2.0 L·min^{-1} of oxygen during jogging.

List two eventual fates of lactic acid:

1.

2.

Many Levels of Steady-Rate

What two metabolic-physiologic factors determine a person's ability to perform exercise at a steady-rate?

1.

2.

Oxygen Deficit

What does the oxygen deficit most likely represent in terms of exercise metabolism?

OXYGEN DEFICIT IN TRAINED AND UNTRAINED

Indicate how the previous graph (page 79) of oxygen uptake during 10 minutes of moderate running would differ for an endurance trained athlete.

Maximal Oxygen Uptake ($\dot{V}O_{2max}$)

Draw and label the oxygen uptake curve during exercise of progressively increasing work intensity (every 3 min) to exhaustion. Indicate the $\dot{V}O_{2max}$ (Hint: Refer to Fig. 5.5 in your textbook.)

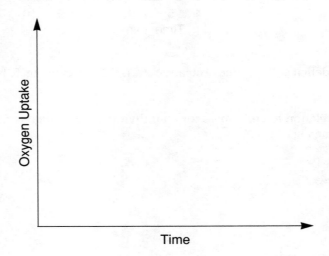

Is it possible to exercise at an intensity greater than the level that elicited $\dot{V}O_{2max}$? Explain.

FAST- AND SLOW-TWITCH MUSCLE FIBERS

Fast-Twitch Fiber

List two characteristics of fast-twitch (type II) muscle fibers

1.

2.

What energy system most often supports fast-twitch muscle fiber activity?

Slow-Twitch Fiber

List two characteristics of slow-twitch or (type I) muscle fibers.

1.

2.

What energy system most often supports slow-twitch muscle fiber activity?

ENERGY SPECTRUM OF EXERCISE

Explain how an understanding of the energy spectrum in exercise can help you to design a better training program for athletes in different sports.

Intensity and Duration Determine the Blend

Explain why energy transfer during exercise should be viewed as a continuum.

What two factors determine which energy system and metabolic mixture predominate used during exercise.

1.

2.

Nutrient-Related Fatigue

List three possible explanations as to why muscle glycogen depletion occurs during prolonged exercise coincides with reduced exercise capacity.

1.

2.

3.

OXYGEN UPTAKE DURING RECOVERY: THE SO-CALLED "OXYGEN DEBT"

Draw and label the oxygen uptake curves during recovery from light, steady-rate exercise and exhaustive exercise. Include the fast the slow component of the recovery where applicable. (Hint: Refer to Fig. 5.7 in your textbook.)

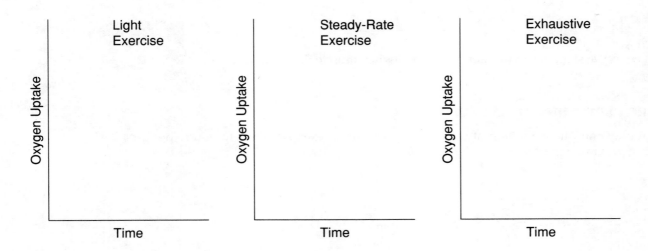

List four reasons why post-exercise oxygen uptake takes longer to return to baseline following strenuous exercise compared to moderate exercise.

1.

2.

3.

4.

Metabolic Dynamics of Recovery Oxygen Uptake

Traditional Concepts: A.V. Hill's Oxygen Debt Theory

Outline A.V. Hill's concepts explaining the recovery oxygen uptake.

What two components did Hill propose to explain the recovery $\dot{V}O_2$?

1.

2.

TESTING HILL'S OXYGEN DEBT THEORY

Describe one way researchers tested Hill's oxygen debt theory. What were the results?

UPDATED THEORY TO EXPLAIN EPOC

Outline the current thinking regarding causes for the excess post-exercise oxygen uptake (EPOC) from steady-rate and non steady-rate exercise.

Steady-rate:

Non steady-rate:

OTHER FACTORS AFFECT EPOC

List four factors believed responsible for "causing" an elevated post exericse oxygen uptake.

1.

2.

3.

4.

IMPLICATIONS OF EPOC FOR EXERCISE AND RECOVERY

List two ways knowledge of EPOC influences strategies for sports training.

1.

2.

Optimal Recovery from Steady-Rate Exercise

What procedures optimize recovery from steady-rate exericse?

Optimal Recovery from Non–Steady-Rate Exercise

What procedures optimize recorvery from steady-rate exericse?

Intermittent Exercise: The Interval Training Approach

What major advantage of intermittent exercise translates to improved performance and training?

Describe a practical application of the concept of intermittent exercise (interval) training. (Hint: refer to Table 5.1).

Practice Quiz

MULTIPLE CHOICE

1. Blood lactate accumulates at what percent of a healthy, untrained person's $\dot{V}O_{2max}$?
   a. 45%
   b. 55%
   c. 65%
   d. 75%
   e. Blood lactate does not accumulate in healthy individuals

2. The oxygen uptake curve during constant-load submaximal exercise:
   a. Increases rapidly during the first two minutes of exercise
   b. Plateaus within the first thirty seconds
   c. Never a plateaus
   d. Never plateaus if fat serves as the exercise fuel
   e. a and d

3. Which is true about oxygen deficit in the trained and untrained?
   a. A trained person reaches a steady-rate more rapidly and thus has a smaller oxygen deficit
   b. An untrained person reaches a steady-rate more rapidly and thus has a smaller oxygen deficit
   c. A trained person reaches a steady-rate more rapidly and thus has a larger oxygen deficit
   d. The trained and untrained reach a steady-rate at the same time during exercise
   e. The untrained always have a smaller oxygen deficit

4. Fast-twitch muscle fibers:
   a. Type I fibers
   b. Possess high capacity for aerobic ATP production
   c. Fatigue resistant
   d. Predominate in short-term, sprint activities
   e. All of the above

5. The recovery oxygen uptake reflects which of the following effects of exercise?
   a. Anaerobic metabolism
   b. Respiratory adjustments
   c. Thermal adjustments
   d. Circulatory adjustments
   e. All of the above

6. Most rapid form of recovery from steady-rate exercise:
   a. Active recovery
   b. Passive recovery
   c. Warm water immersion
   d. Cold water immersion
   e. b and d

7. The most effective way to train the aerobic energy system:
   a. Sprint training
   b. Strength training
   c. "Super-maximal" exercise with brief rest intervals
   d. Long-distance training
   e. Combination of a and b

8. The best indicator of one's ability to sustain high-intensity aerobic exercise is the:
   a. $\dot{V}O_{2max}$
   b. Maximal steady-rate
   c. Maximal oxygen debt
   d. Maximal oxygen deficit
   e. Maximal ventilation volume

9. Alactacid debt:
   a. Represents the fast portion of recovery oxygen uptake curve
   b. Represents the slow portion of recovery oxygen uptake curve
   c. Indicates the quantity of protein used as fuel
   d. "Pays back" the oxygen deficit
   e. Converts fatty acids to glycogen

10. Success in 220-yd sprint running and weight lifting requires a highly trained:
   a. Aerobic energy system
   b. Lactic acid energy system
   c. ATP-PCr energy system
   d. b and c
   e. a and c

T R U E   /   F A L S E

1. _____ Specific anaerobic training can increase ability to generate high lactic acid levels.

2. _____ The oxygen debt represents the difference between total oxygen consumed during exercise and the amount that would have been consumed had steady-rate occurred instantaneously.

3. _____ Recruitment of slow-twitch muscle fibers favors lactic acid formation.

4. _____ The point when the oxygen uptake plateaus with increasing exercise intensity represents the maximal aerobic steady-rate or point of OBLA.

5. _____ Energy transfer mechanisms exists on a continuum in different intensities of exercise.

6. _____ The oxygen deficit describes the oxygen consumed in recovery from exercise.

7. _____ Active aerobic exercise accelerates lactic acid removal during recovery.

8. _____ Oxygen uptake during most forms of exercise does not relate to body size.

9. _____ OBLA occurs at a higher percentage of an untrained person's $\dot{V}O_{2max}$ than for a trained person.

10. _____ Slow-twitch muscle fibers have the capacity to generate more total ATP than fast-twitch fibers.

CROSSWORD PUZZLE

ACROSS

CHAPTER 5

3. reduced training volume before competition
7. these twitch fibers activated in non steady rate exercise
9. rapid phase of recovery from exercise is the _____ portion of the "oxygen debt"
10. _____ recovery usually involves inactivity after activity
12. these twitch fibers activated in steady rate exercise
14. point of abrupt increase in blood lactate
20. the _____ term energy system powers all-out exercise for about 90 s
21. method of performing exercise with rest periods (abbr.)
23. acronym for recovery oxygen uptake
25. approximately 5 mMol of ATP and 15 mMol of CP are _____ within each kg of muscle
26. _____ recovery is often called "cooling down" or "tapering off"
27. high _____ phosphates provide the immediate energy for exercise
28. devise to measure work

DOWN

1. the buffered form of lactic acid
2. difference between total oxygen consumed during exercise and amount that would have been consumed had a steady-rate metabolism occurred
4. marker enzyme for glycolytic capacity
5. slow component of "oxygen debt"
6. phosphocreatine (abbr)
8. the lactate _____ is that point in exercise where lactate rises
9. Nobel physiologist _____ Vivian Hill
11. type of training involving rest:relief segments
13. aerobic capacity (abbr.)
15. movement of lactate between cells enables glycogenolysis in one cell to supply other cells with fuel for oxidation
16. these phosphates power all-out exercise for 6–8 s
17. steady _____ is where exercise energy requirements are met by aerobic resynthesis of ATP
18. this energy system involves aerobic metabolism
19. this energy system relies on high energy phosphates
22. final electron acceptor in energy metabolism
24. onset of blood lactic acid buildup (abbr.)

Measurement of Human Energy Expenditure

Define Key Terms and Concepts

1. Atwater-Rosa human calorimeter

2. August Krogh

3. Calorific value for oxygen

4. Closed-circuit spirometry

5. Direct calorimetry

6. Douglas bag technique

7. Haldane

8. Indirect calorimetry

9. Micro-Scholander

10. Nonprotein RQ

11. Open-circuit spirometry

12. Portable spirometer

13. Respiratory exchange ratio (RER)

14. Respiratory quotient (RQ)

Study Questions

HEAT PRODUCED BY THE BODY

List two methods to determine heat production by the body.

1.

2.

Direct Calorimetry

Describe direct calorimetry to measure human heat production.

Indirect Calorimetry

List the two methods of indirect calorimetry.

1.

2.

Closed-Circuit Spirometry

Give one disadvantage of closed-circuit spirometry during exercise studies

Open-Circuit Spirometry

List one positive aspect of each of the following procedures of indirect calorimetry during exercise studies.

Portable Spirometry:

Bag Technique:

Computerized Instrumentation:

PORTABLE SPIROMETRY

Who were the first scientists to use portable spirometry?

BAG TECHNIQUE

What kind of breathing valve must be used with the bag technique for open-circuit spirometry?

COMPUTERIZED INSTRUMENTATION

List three instruments that need to be interfaced with a computer for on-line measurement of oxygen uptake.

1.

2.

3.

CALIBRATION METHODS

Name two methods for analyzing gas mixtures commonly used for calibrating electronic gas analyzers.

1.

2.

DIRECT VERSUS INDIRECT CALORIMETRY

Give an example of the degree of agreement between energy expenditure obtained by direct and indirect calorimetry.

Caloric Transformation For Oxygen

Assuming combustion of a mixed diet, give the rounded value for the number of calories released per liter of oxygen consumed?

RESPIRATORY QUOTIENT (RQ)

Write the formula for the RQ.

RQ for Carbohydrate

What is the RQ for carbohydrate?

RQ for Lipid

What is the RQ for lipid?

RQ for Protein

What is the RQ for protein?

RQ for a Mixed Diet

Give the RQ for a diet of approximately 40% carbohydrate and 60% lipid. What is the corresponding caloric equivalent?

RQ:

Caloric equivalent:

RESPIRATORY EXCHANGE RATIO (R)

Discuss the difference between RQ and R?

How can exhaustive exercise cause the R to increase above 1.00?

What could cause the R to be less than 0.70?

Practice Quiz

MULTIPLE CHOICE

1. Which is true about direct and indirect calorimetry?
   a. Closed- and open-circuit spirometry are methods of direct calorimetry
   b. A human calorimeter is used for indirect calorimetry
   c. Indirect calorimetry is highly accurate and much less expensive than direct calorimetry
   d. Direct calorimetry measures $\dot{V}O_2$ and $\dot{V}CO_2$ to estimate energy expenditure
   e. a and d

2. Approximately _____ kcal of heat energy liberates when carbohydrate oxidizes to CO_2 and H_2O in 1 liter of oxygen:
   a. 4
   b. 5
   c. 6
   d. 7
   e. None of the above

3. Which is false about RQ?
   a. RQ for carbohydrate equals 1.00.
   b. RQ for lipid equals 0.90.
   c. RQ for protein equals 0.82.
   d. RQ for a mixed diet ranges between 0.70 and 1.00.
   e. All of the above

4. Energy metabolism ultimately depends on:
   a. Oxygen utilization
   b. STP regeneration
   c. SDA availability
   d. Carbohydrate availability
   e. None of the above

5. Spirometry represents two types of indirect calorimetry:
   a. Nitrogen analysis, oxygen analysis
   b. Closed-circuit, direct
   c. Open-circuit, Douglas bag
   d. Bomb calorimetry, closed-circuit
   e. Closed-circuit, open-circuit

6. Closed-circuit spirometry:
   a. The person rebreathes CO_2 with O_2 being absorbed
   b. The respiratory exchange ratio measures oxygen uptake
   c. CO_2 in exhaled air is absorbed by soda lime
   d. The respiratory exchange ratio is always determined
   e. None of the above

7. In open-circuit spirometry the subject:
   a. Inhales ambient air
   b. Rebreathes from a spirometer containing pure oxygen
   c. Inhales pure oxygen from a Douglas bag
   d. Is always fully hydrated
   e. Inhales and exhales approximately 100 times every 30 s

8. To estimate the body's heat production one need only know the:
   a. RQ
   b. RQ and amount of oxygen consumed
   c. RQ and amount carbon dioxide produced
   d. Oxygen consumed
   e. None of the above

9. Non-protein RQ:
   a. Is only used during rest
   b. Assumes that protein contributes 25% to total energy production
   c. Assumes only a small contribution of protein to energy metabolism
   d. Is only used during high-intensity anaerobic exercise
   e. None of the above

10. The RQ for a mixed diet:
    a. Assumes combustion of 12% protein, 35% carbohydrate, and 25% lipid
    b. Assumes combustion of 40% carbohydrate and 60% lipid
    c. Equals the RQ for a high-carbohydrate diet at the same oxygen uptake level
    d. Assumes the combustion of an equal mixture of lipid and protein
    e. Depends on the nutrient mixture's digestibility coefficient

TRUE / FALSE

1. _____ The most common technique to measure oxygen uptake is the open-circuit method.

2. _____ Portable spirometry, bag technique, and computerized instrumentation are examples of direct calorimetry.

3. _____ The Weir Factor measures the ratio of metabolic gas exchange.

4. _____ The ratio of CO_2 produced to O_2 consumed equals the respiratory quotient.

5. _____ RQ for lipid equals 0.70.

6. _____ For any R value, a correspondingly calorific value exists for each liter of oxygen consumed.

7. _____ RQ for a mixed diet equals 0.82.

8. _____ RQ rises above 1.0 when the body converts excess dietary lipids to glycogen.

9. _____ R values in excess of 1.00 reflect excess CO_2 "blow-off".

10. _____ The portable spirometer weighs about 12 pounds.

CROSSWORD PUZZLE

ACROSS

2. this ventilatory maneuver increases the RQ
4. scientist who built and perfected the first human calorimeter of major scientific importance
5. RQ is 0.70 for this macronutrient
8. 4.84 kcal liberated when mixed diet is burned in this many liters of O2
9. measurement of heat production
12. CO2 + O2 in exhaustive exercise is called the _____ ratio
16. CO2 + O2
17. calorimetry that infers energy from change in water temperature
18. oxygen uptake indirectly infers this

DOWN

1. another term for KOH
3. Kcal equivalent For 1 L Oxygen assuming the combustion of a mixed diet
6. collection device in open circuit spirometry
7. general RQ value for this macronutrient in 0.82
9. spirometry where subject rebreathes from a prefilled container
10. recovery from this type of exercise associated with RQ below 0.70

CHAPTER 6

11. simple method for estimating caloric expenditure from measures of pulmonary ventilation and expired oxygen percent; Scottish physician - physiologist
13. calorimetry that infers energy from O2 uptake
14. douglas bag method is example of this type of spirometry
15. scientists in this country developed the portable spirometer

Energy Expenditure During Rest and Physical Activity

1. Basal metabolic rate (BMR)

2. Body surface area (BSA)

3. Buoyancy effects in swimming

4. Dietary-induced thermogenesis

5. Drafting

6. Drag forces

7. Economy of movement

8. Gross energy expenditure

9. Heavy work

10. Indirect calorimetry

11. Light work

12. Maximal work

13. Mechanical efficiency

14. MET

15. Negative work

16. Net energy expenditure

17. One $kcal \cdot kg^{-1} \cdot km^{-1}$

18. Portable spirometer

19. RDEE

20. Skin friction drag

21. Steady-rate

22. Swimming flume

23. TDEE

26. Weight bearing exercise

24. Viscous pressure drag

27. Weight supported exercise

25. Wave drag

Study Questions

ENERGY EXPENDITURE AT REST

List three factors that influence 24-hr energy expenditure.

1.

2.

3.

ENERGY EXPENDITURE AT REST: BASAL METABOLIC RATE

List three standardized conditions for measuring BMR.

1.

2.

3.

Give the average BMR for a college-age man and woman.

Man:

Woman:

INFLUENCE OF BODY SIZE ON RESTING METABOLISM

Why is a women's BMR at any age lower compared with men?

Give the preferred way to express BMR to account for body size differences?

Estimate the BMR for a 21-year-old female who weighs 50 kg and is 66 inches in height.

ESTIMATING RESTING DAILY ENERGY EXPENDITURE (RDEE)

Determine the 24-hr RDEE for a person with a BMR of 35 kcal·m^2·hr^{-1} and a surface area of 1.5 m^2.

FACTORS AFFECTING ENERGY EXPENDITURE

List three factors that affect a person's total daily energy expenditure.

1.

2.

3.

Physical Activity

Physical activity normally accounts for what percent of a person's TDEE?

Dietary-Induced Thermogenesis

Dietary-induced thermogenesis reaches a maximum how long after a meal?

List two factors that affect the magnitude of dietary-induced thermogenesis.

1.

2.

Climate

What two physiologic effects from exercising in the heat could elevate metabolism during exercise?

1.

2.

Pregnancy

List two ways that women can protect their fetus during exercise.

1.

2.

List two maternal physiologic adaptations to pregnancy.

1.

2.

ENERGY EXPENDITURE DURING PHYSICAL ACTIVITY

Compute the approximate total kcal expenditure if a rower averages a $\dot{V}O_2$ of 2.0 L·min^{-1} for 30 minutes of rowing.

ENERGY COST OF RECREATIONAL AND SPORT ACTIVITIES

Effect of Body Mass

Display the relationship between body mass and energy expenditure during weight-bearing exercise. (Hint: Refer to Fig. 7.3 in your textbook).

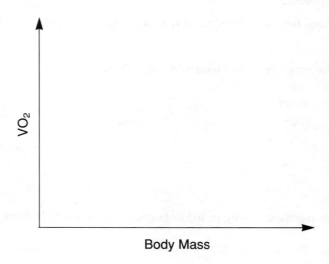

AVERAGE DAILY RATES OF ENERGY EXPENDITURE

Give the average daily caloric expenditure for men and women between ages 23 and 50 years.

Men:

Women:

Give the equation to compute the total caloric cost of a physical activity:

Total caloric cost = _____ × _____

CLASSIFICATION OF WORK

List two factors that affect how one rates the difficulty of a particular physical task.

1.

2.

Describe the physical activity ratio (PAR).

The MET

One MET equals approximately _____ L·min^{-1}.

One MET equals approximztely _____ mL·kg^{-1}·min^{-1}.

If body mass equals 70 kg, and max MET level equals 12, estimate maximum oxygen uptake?

Heart Rate to Estimate Energy Expenditure

Heart rate relates linearly to what other factor during physical activity?

List two limitations in using heart rate to estimate $\dot{V}O_2$ (energy expenditure) during exercise.

1.

2.

List three factors (other than oxygen uptake) that influence exercise heart rate.

1.

2.

3.

ENERGY EXPENDITURE DURING WALKING, RUNNING, AND SWIMMING

What primary factor accounts for differences among people in daily energy expenditure?

ECONOMY OF MOVEMENT

Is running economy of a trained athlete higher or lower than the economy of an untrained person in the same exercise?

Exercise Oxygen Uptake Reflects Economy

True or False: Trained runners show a greater running economy than untrained runners because they have a lower $\dot{V}O_2$ at the same running speed.

True or False: Type I muscle fibers exhibit greater mechanical efficiency than type II fibers.

MECHANICAL EFFICIENCY

Write the formula for calculating mechanical efficiency (%).

List two factors that affect the mechanical efficiency of human movement.

1.

2.

Running Economy: Children and Adults, Trained and Untrained

Give the magnitude of the difference in running economy of children compared with adults.

ENERGY EXPENDITURE DURING WALKING

Draw and label the relationship between speed of horizontal walking and oxygen uptake. (Hint: Refer to Fig. 7.8 in your textbook.)

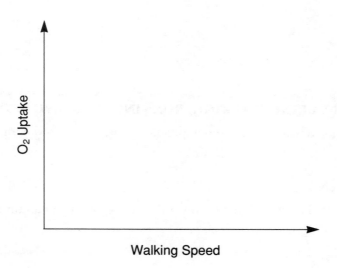

Competition Walking

Describe differences in economy between walking and running at speeds faster than 8 km·h^{-1}.

Effects of Body Mass

Calculate total energy expenditure (kcal) for a person who weighs 54 kg and walks at 4.83 km·hr^{-1} for 20 minutes. (Hint: Refer to Table 7.6 in your textbook.)

Effects of Terrain and Walking Surface

Which of the walking surfaces discussed in the text results in the highest energy expenditure at a given submaximum speed?

Effects of Footwear

What differences exist in energy expenditure for carrying weight on the torso, feet, or ankles?

Torso:

Feet:

Ankles:

Use of Hand-Held and Ankle Weights

How do ankle weights affect the energy cost of walking?

Discuss the desirability of wearing ankle weights during running.

Give the difference in impact force on the lower extremities during running compared to walking.

ENERGY EXPENDITURE DURING RUNNING

From an energy standpoint, at what speed does it become economical to begin to run?

Draw and label the relationship between the speed of horizontal running and energy expenditure. (Hint: Refer to Fig. 7.8 in your textbook.)

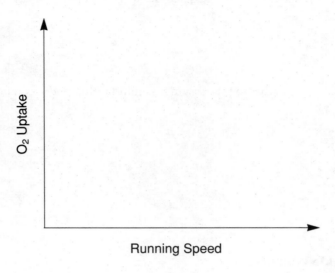

Economy of Running

What effect does speed have on the energy cost of running a given distance?

Give the net energy cost of horizontal running per kilogram of body mass per kilometer of distance?

Energy Cost Values

True or false? Running speed does not affect the energy requirement of running a given distance.

Stride Length, Stride Frequency, and Running Speed

In terms of stride frequency and length, running speed increases mainly by _____

_____.

Optimum Stride Length

What effect does shortening normal running stride and increasing the number of steps to maintain speed have on running economy?

Effects of Air Resistance

What three factors influence the effect of air resistance on energy cost of running?

1.

2.

3.

Drafting

Quantify the performance and economy benefits of drafting.

Treadmill Versus Track Running

Give the difference in energy cost between running on a track and a motorized treadmill.

Marathon Running

World class marathon runners run at about what percent of their $\dot{V}O_{2max}$?

A 70 kg person expends how many kcal to run a 26.2 mile marathon. (Hint: Refer to Table 7.8 in your textbook.)

ENERGY EXPENDITURE DURING SWIMMING

List two factors that contribute to differences in energy expenditure between swimming and walking and running.

1.

2.

Methods of Measurement

Describe two methods for measuring energy expenditure during swimming.

1.

2.

Energy Cost and Drag

List the three components of the total drag force during swimming.

1.

2.

3.

Energy Cost, Swimming Velocity, and Skill

How does steady-rate oxygen uptake at a given velocity of swimming differ between trained and untrained swimmers?

Effects of Buoyancy: Men Versus Women

Why can women achieve a lower energy expenditure than men while swimming at the same speed?

Give the percent difference in energy cost of swimming between males and females?

Practice Quiz

MULTIPLE CHOICE

1. Aerobic requirements of submaximum running (up to 286 m·min^{-1}) on the treadmill or track (either on level or up a grade):
   a. Greater on a treadmill
   b. Greater on a track
   c. Same on a track and treadmill
   d. Lower on a treadmill
   e. None of the above

2. Three components of total drag force during swimming:
   a. Wave, skin, viscous
   b. Wave, temperature, pressure
   c. Viscous, pressure, hydraulic
   d. Hydraulic, pressure, internal
   e. Wave, hydraulic, skin

3. Overcoming air resistance accounts for between ____ to ____ percent of the energy cost of running in calm weather:
   a. 10, 15
   b. 3, 9
   c. 20, 30
   d. 1, 3
   e. None of the above

4. Which exerts the greatest influence on total daily energy expenditure?
   a. Percent body fat
   b. Resting metabolic rate
   c. Thermogenic influence of food consumed
   d. Energy expended during physical activity and recovery
   e. Factors affect an equal affect

5. Evaluate _____ during steady-rate exercise to establish differences among individuals in economy of physical effort:
   a. Heart rate
   b. Carbon dioxide production
   c. Oxygen uptake
   d. Blood pressure
   e. All of the above

6. Which is true about running economy?
   a. Boys and girls are less economical in running compared to adults
   b. Boys and girls are more economical in running compared to adults
   c. At the same speed, trained runners run at a lower oxygen uptake than untrained runners
   d. At the same speed trained runners run at a higher oxygen uptake than untrained runners
   e. a and c

7. Which is false about running?
   a. Net energy cost for horizontal running per kg body mass per km equals approximately 1 kcal
   b. Linear relationship exists between oxygen uptake and running speed at normal running speeds
   c. More economical to run than walk at speeds greater than $8 \text{ km} \cdot \text{h}^{-1}$
   d. Hand or ankle weights decrease energy expenditure during running
   e. a and d

8. Which of the following best defines 1 "Met":
   a. Maximum energy expenditure of training
   b. Minimum resting energy expenditure
   c. A unit of resting metabolism
   d. Energy metabolism in training
   e. Equivalent of one PAR

9. Elite swimmers swim a particular stroke at a given velocity at a _____ $\dot{V}O_2$ than untrained swimmers:
   a. Lower
   b. Higher
   c. 5 to 10% higher
   d. 20 to 30% higher
   e. None of the above

10. Energy cost of cutting through a headwind:
    a. Counterbalanced on one's return with the tailwind
    b. The same as the energy cost of the tailwind
    c. Less than the increase in the oxygen cost of drafting
    d. Greater than the reduction in exercise $\dot{V}O2$ observed with an equivalent wind velocity at one's back
    e. None of the above

TRUE / FALSE

1. _____ The total caloric cost of running a given distance at a steady-rate oxygen uptake is about the same fast or slow.

2. _____ The net energy cost for horizontal running equals 1 kcal ($1 \text{ kcal} \cdot \text{kg}^{-1} \cdot \text{km}^{-1}$).

3. _____ Except at rapid speeds, running speed increases mainly by increasing stride frequency.

4. _____ One "best" running style characterizes elite runners.

5. _____ While running, children require significantly more oxygen to transport their body mass compared to adults.

6. _____ Heavier people expend more energy to perform the same activity than lighter people.

7. _____ The economy of walking faster than $5 \text{ km} \cdot \text{h}^{-1}$ equals one-half that of running at similar speeds.

8. _____ The energy cost of carrying weights on the feet or ankles is less than similar weight on the torso.

9. _____ The energy cost of swimming a given distance is four times greater than running the same distance.

10. _____ Untrained swimmers perform a particular stroke at a lower $\dot{V}O_2$ than trained swimmers.

CROSSWORD PUZZLE

ACROSS

2. these individuals often have a blunted thermic response to eating
4. the _____ between the economy of walking and running occurs at about 8 km per hour (5.0 mph)
5. differences in _____ partly explains women's greater swimming economy
8. work requiring nine times or more the resting metabolism
11. maternal "hyperventilation" in exercise has been attributed to the effect of this hormone
16. older term for increased metabolism with eating (abbr.)
17. body mass minus fat mass (abbr.)
19. stimulating effect of food ingestion on metabolism (abbr.)
21. this macronutrient elicits the highest thermic effect
22. total energy expended during a day (abbr.)
23. term meaning "heat producing"
24. has about a 5–20% effect on metabolism
25. this bodily response to cold stress can easily double or triple the metabolic rate

DOWN

1. physiologic variable directly related to oxygen uptake throughout a large portion of the aerobic work range
3. a lowering of this blood macronutrient could adversely affect fetal development
5. external surface of the body (abbr.)
6. minimum energy required to sustain the body's vital functions in the waking state (abbr.)
7. work requiring up to three times resting metabolism

CHAPTER 7

9. work requiring six to eight times the resting metabolism
10. _____ absorptive state is when no food is eaten for at least 12 hrs
11. physical activity classification (abbr.)
12. component of thermogenesis related to specific food ingested
13. steady-state oxygen uptake
14. mechanical work accomplished ÷ input of energy × 100
15. metabolism that is slightly higher than BMR (abbr.)
18. variations in body _____ largely explain the gender difference in BMR
20. multiple of the resting metabolic rate (abbr.)

Evaluating Energy-Generating Capacities During Exercise

Define Key Terms and Concepts

1. Absolute $\dot{V}O_{2max}$

2. Anaerobic capacity

3. Anaerobic power

4. Average power

5. Balke protocol

6. Bruce protocol

7. Genotype

8. Glycogen depletion

9. Glycolytic power

10. Graded exercise

11. HR – $\dot{V}O_2$ line

12. Individual differences

13. Joule

14. Jumping power tests

15. Katch test

16. Maximal lactate levels

17. $\dot{V}O_{2peak}$

18. Peak power

19. Phenotype

20. Power

21. Quebec 10-s test

22. Queens College step test

23. Ramp test

24. Rate of fatigue

25. Relative $\dot{V}O_{2max}$

26. Stair sprinting power test

27. "True" $\dot{V}O_{2max}$

28. Watt

29. Wingate test

Study Questions

OVERVIEW OF ENERGY TRANSFER CAPACITY DURING EXERCISE

Identify the energy system that primarily supports each of the following activities:

Standing vertical jump and reach

Four hundred meter run

Three mile run

Power = _____ × _____ ÷ _____ .

ANAEROBIC ENERGY: IMMEDIATE AND SHORT-TERM ENERGY SYSTEMS

Evaluation of the Immediate Energy System

What type of tests typically measure the immediate anaerobic energy system?

Write the units of measurement for power tests.

Stair-Sprinting Power Test

Compute the power output for a person who weighs 55 kg and traverses nine steps in 0.75 seconds (vertical rise of each step equals 0.175 meters).

Jumping-Power Test

Give one factor that might limit power jumping test scores as measures of the power output capacity of intramuscular high-energy phosphates.

Other Immediate Energy Power Tests

List two tests (other than stair-sprinting) to estimate power output capacity of the immediate energy system.

1.

2.

Relationships Among Power Tests

Explain why the relationship among power tests is not consistently strong. Give an example.

Evaluation of the Short-Term Energy System

What type of test best estimates the power output capacity of the short-term energy system?

Would tests of the short-term energy system exhibit task specificity? Explain.

Performance Tests of Glycolytic Power

List two tests to measure short-term energy transfer capacity.

1.

2.

List three factors that affect anaerobic power performance.

1.

2.

3.

The "Katch Test" and the "Wingate Test" measure what aspect of energy-generating capacity?

Explain the major difference between "peak power" and "average power."

Other Anaerobic Tests

Name one other exercise performance test to measure anaerobic power and capacity.

LOWER IN CHILDREN

List one possible reason children perform poorer on tests of short-term anaerobic power compared with adolescents and adults.

GENDER DIFFERENCES

Do differences in body composition fully explain the gender differences in anaerobic power capacity? Discuss.

Blood Lactate Levels

Draw and label the relationship between blood lactate levels and the duration of an all-out exercise test of short duration. (Hint: Refer to Fig. 8.5 in your textbook.)

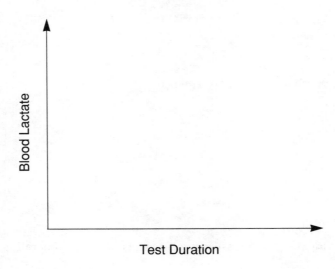

Glycogen Depletion

List two factors that determine muscle glycogen depletion in different muscle fiber types within the same muscle.

1.

2.

Individual Differences in Anaerobic Energy Transfer Capacity

List three factors that differ among individuals in their capacity to generate short-term energy.

1.

2.

3.

Effects of Training

How does training affect the depletion pattern of anaerobic substrates and capacity for lactic acid accumulation? (Hint: Refer to Fig. 8.7 in your textbook.)?

Anaerobic substrate depletion

Lactic acid accumulation

Buffering of Acid Metabolites

Do athletes have an enhanced buffering capacity compared to non-athletes? Explain.

Motivation

What relationship would you expect between "pain tolerance" and one's capacity for anaerobic exercise? Explain.

AEROBIC ENERGY: LONG-TERM ENERGY SYSTEM

List three categories of athletes that typically exhibit high values for $\dot{V}O_{2max}$?

1.

2.

3.

Maximal Oxygen Uptake Measurement

Describe the "gold standard" to indicate attainment of true $\dot{V}O_{2max}$.

Criteria for $\dot{V}O_{2max}$.

List three additional criteria for establishing $\dot{V}O_{2max}$.

1.

2.

3.

Tests of Aerobic Power

List two general criteria for a good $\dot{V}O_{2max}$ test.

1.

2.

Comparison Between $\dot{V}O_{2max}$ Tests

List two types of graded exercise tests to elicit $\dot{V}O_{2max}$.

1.

2.

Commonly Used Treadmill Protocols

List three factors common to all treadmill tests of $\dot{V}O_{2max}$.

1.

2.

3.

Manipulating Test Protocol to Increase $\dot{V}O_{2max}$

Administering two successive $\dot{V}O_{2max}$ tests to the same person (2-min recovery between tests) produces what effect on test scores? Explain.

Factors Affecting Maximal Oxygen Uptake

List six factors that influence $\dot{V}O_{2max}$.

1.   4.

2.   5.

3.   6.

Exercise Mode

Indicate the most common piece of exercise equipment to determine $\dot{V}O_{2max}$.

Heredity

Give the estimated magnitude of heredity in determining $\dot{V}O_{2max}$.

Training State

Give the general range for $\dot{V}O_{2max}$ improvement with training.

Gender

Give two reasons for gender-related differences in $\dot{V}O_{2max}$.

1.

2.

Body Composition

Differences in body mass explain ———— % of the differences in $\dot{V}O_{2max}$ ($L \cdot min^{-1}$) among individuals.

Explain the "best" way to express $\dot{V}O_{2max}$.

ARGUMENT FOR BIOLOGICAL DIFFERENCES BETWEEN MEN AND WOMEN

List two points that support the existence of real biological differences in $\dot{V}O_{2max}$ between sexes.

1.

2.

Age

After age 25, $\dot{V}O_{2max}$ generally decreases approximately what percent each year?

ABSOLUTE VALUES/RELATIVE VALUES

Discuss the trend in $\dot{V}O_{2max}$ in absolute and relative terms between the ages 6 and 16 years in boys and girls.

MAXIMAL OXYGEN UPTAKE PREDICTIONS

List two popular methods to predict $\dot{V}O_{2max}$ from submaximal exercise.

1.

2.

A Word of Caution About Predictions

Give the average standard error for predicting $\dot{V}O_{2max}$.

Heart Rate Predictions of $\dot{V}O_{2max}$

What four assumptions underlie the use of heart rate to predict $\dot{V}O_{2max}$?

1.   3.

2.   4.

Practice Quiz

MULTIPLE CHOICE

1. After age 25, the $\dot{V}O_{2max}$ declines approximately ____ % per year:
   a. 1
   b. 3
   c. 5
   d. 10
   e. No decrease

2. During exercise to approximately 80% of maximum intensity, $\dot{V}O_2$ and heart rate:
   a. Relate inversely proportional
   b. Relate linearly
   c. Do not relate
   d. Relate curvilinearly
   e. Relate inversely to cardiac output

3. Which is not a necessary assumption for accurately predicting $\dot{V}O_{2max}$ from exercise heart rate:
   a. Linearity of the HR-$\dot{V}O_2$ relationship
   b. Similar maximal HR for all subjects of the same age
   c. No day-to-day variation in exercise HR
   d. Significant gender differences in maximal HR
   e. c and d

4. All of the following except _____ affect differences among individuals in short-term energy capacity:
   a. Training
   b. Motivation
   c. Distribution of fast-twitch fibers
   d. Buffering of acid metabolites
   e. Aerobic capacity

5. Which energy system predominantly powers 60- to 90-seconds of high-intensity, all-out exercise:
   a. Immediate system
   b. Short-term system
   c. Long-term system
   d. ATP-PCr energy system
   e. All contribute equally

6. Which represents the main criterion for $\dot{V}O_{2max}$:
   a. When $\dot{V}O_2$ fails to increase with increasing exercise intensity
   b. Blood lactic acid levels reach 70 or 80 mg per 100 mL of blood or higher
   c. Attainment of near age-predicted maximum heart rate
   d. Respiratory exchange ratio exceeds 1.0.
   e. None of the above

7. Which is true about the relative $\dot{V}O_{2max}$ values for men and women:
   a. Values for women average 15-30% below values for men
   b. Values for women average 15-30% above values for men
   c. Values for women average 0-15% below values for men
   d. Values for women average 0-15% above values for men
   e. None of the above

8. The relative contribution of anaerobic and aerobic energy transfer largely depends on:
   a. Intensity of exercise
   b. Duration of exercise
   c. Mode of exercise
   d. a and b
   e. None of the above

9. The Wingate test cannot evaluate:
   a. Peak power output
   b. Average power output
   c. Maximal oxygen uptake
   d. Rate of fatigue
   e. a and c

10. The stair-sprinting power test evaluates this energy system:
    a. Immediate system
    b. Short-term system
    c. Long-term system
    d. a and c
    e. All of the above

TRUE / FALSE

1. _____ Men tend to have a higher $\dot{V}O_{2max}$ than women because of their higher hemoglobin levels.

2. _____ Improvements in aerobic capacity with training usually range between 30 and 65%.

3. _____ Peak power represents total power generated during a 30-s, all-out exercise period.

4. _____ Peak $\dot{V}O_2$ is usually close in value to the actual $\dot{V}O_{2max}$.

5. _____ Jumping and stair-stepping tests are frequently used to predict maximal $\dot{V}O_{2max}$.

6. _____ More specificity than generality exists for each energy system.

7. _____ Glycogen depletion occurs selectively in fast- and slow-twitch fibers depending on exercise intensity and duration.

8. _____ Altering acid-base balance in the direction of alkalosis prior to exercising can enhance short-term, high-intensity exercise performance.

9. _____ Heredity exerts only a small effect on an individual's $\dot{V}O_{2max}$.

10. _____ Differences in body mass explain about seventy percent of the differences in absolute $\dot{V}O_{2max}$ ($L \cdot min^{-1}$) among individuals.

CROSSWORD PUZZLE

CHAPTER 8

ACROSS

2. graded exercise stress test (abbr.)
3. highest VO2 in maximal testing
4. energy per unit time generated by immediate and short term energy systems
5. vertical _____ test often used to measure immediate energy system
7. SI unit equal to 1.0 j·s − 1
12. term describing deviations of individuals from the average or from each other
14. work per unit time
15. total energy generated by the immediate and short term energy system
16. type of GXT where the workrate is progressively increased each minute
17. type of test that measures anaerobic power using flight of stairs
18. observable characteristics determined by both genotype and environment

DOWN

1. highest power output achieved during an all out anaerobic test
2. genetic constitution of an individual
5. SI unit for expressing energy
6. Level of blood chemical to assess anaerobic capacity
8. college for which step test is named
9. all-out 30 s test (named for university) on the cycle ergometer with resistance equal to 0.075 kg per kg body mass
10. type of GXT protocol
11. type of GXT protocol
13. physician who pioneered work in "aerobics"; (12 min run test bears his name)

Optimal Nutrition for Exercise and Sports

1. Carbohydrate loading

2. Daily reference values (DRV)

3. Daily values (DV)

4. Diet Quality Index

5. Dietary Guidelines for Americans

6. Eat more, weigh less

7. Food Guide Pyramid

8. Fructose

9. Glucose polymers

10. Glycemic index

11. Ideal oral rehydration solution

12. Ideal sports drink

13. Insulin overshoot response

14. Liquid meals

15. Modified loading procedure

16. Optimal carbohydrate intake

17. Optimal diet

18. Optimal lipid intake

19. Optimal protein intake

20. Osmolality

21. Precompetition meal

22. Prudent diet

23. Reference Daily Intakes (RDI)

25. Ultramarathoner Kouros

24. Tour de France

Study Questions

NUTRIENT REQUIREMENTS

On average, are the diets of endurance athletes much different from sedentary individuals? Explain. (Hint: refer to Table 9.1 in your textbook.)

Recommended Nutrient Intake

Give the average daily kcal intake for a typical American college-age male and female.

Male:

Female:

A prudent diet for a physically active person should provide what percent of total calories from carbohydrates?

Does any health risk exist from subsisting entirely on highly diverse, fiber-rich, complex carbohydrates?

List three dietary guidelines for Americans.

1.

2.

3.

Food Guide Pyramid

Outline the primary focus of the Food Guide Pyramid recommendations for food types.

Differentiate between daily reference values and reference daily intakes.

Diet Quality Index

List three "ideal" dietary goals of the Diet Quality Index.

1.

2.

3.

EXERCISE AND FOOD INTAKE

Do men consume more or less calories (how many?) than females of the same age?

Physical Activity Makes a Difference

Estimate the average "extra" daily caloric expenditure for distance runners who train 100 miles per week.

Extreme Energy Intake and Expenditure: The Tour de France

Give the approximate daily caloric intake for competitors in the Tour de France.

Ultraendurance Running Competition

Based on available data, estimate daily caloric requirements for ultraendurance runners during competition.

Describe the athletic accomplishments of Kouros.

Some Athletes Require Supplementation

Which athletes are prime candidates for needing nutritional supplementation?

Indicate nutrients most likely deficient in these athletes' diets.

Eat More, Weigh Less

How can physically active people actually eat more than the average person, yet weigh less?

THE PRECOMPETITION MEAL

Give the main purpose of the precompetition meal.

Explain whether it is beneficial for an athlete to fast prior to competition.

High Protein: Not the Best Choice

List two reasons why it is important to reduce the protein content of the precompetition meal.

1.

2.

Give one reason why the precompetition meal should consist mainly of carbohydrate.

Ideal Precompetition Meal

List two specific recommendations concerning carbohydrate content and timing of the ideal precompetition meal.

1.

2.

LIQUID MEALS

List three benefits of a liquid precompetition meal.

1.

2.

3.

CARBOHYDRATE INTAKE BEFORE, DURING, AND AFTER INTENSE EXERCISE

What types of exercise severely stress muscle and liver glycogen reserves?

Before Exercise

What two possible negative effects might the intake of high-glycemic carbohydrate have if consumed within 1 hour before exercising?

1.

2.

Describe the "insulin overshoot" phenomenon.

Pre-Exercise Fructose

Even though the gut absorbs fructose more slowly than glucose or sucrose and does not cause an insulin overshoot, why does it *not* qualify as a preferred pre-exercise energy nutrient?

Glycemic Index

What factors determine the glycemic index for a given carbohydrate-containing food?

Glycemic Index and Pre-Exercise Feedings

How can the glycemic index help formulate the immediate pre-exercise feeding?

During Exercise

What is the optimum amount of liquid or solid carbohydrates that should be consumed each hour during high-intensity, long-duration exercise?

GLUCOSE INTAKE, ELECTROLYTES, AND WATER UPTAKE

Why does no insulin overshoot occur when consuming simple sugars during exercise?

Practical Recommendation

Give your recommendation for ingesting fluid prior to and during exercise to optimize fluid replacement.

Before exercise:

During exercise:

Consider Fluid Concentration

List two characteristics of the best rehydration solution.

1.

2.

What is a glucose polymer, and what benefit does it provide for fluid replacement.

Glucose polymer:

Benefit:

Sodium's Potential Benefit

In what way does adding a moderate amount of sodium to ingested fluid benefit the athlete during exercise.

RECOMMENDED ORAL REHYDRATION BEVERAGE

List five qualities of the ideal oral rehydration solution.

1.   4.

2.   5.

3.

Calculate the carbohydrate percentage of a drink that contains 40g of carbohydrate in 500 mL of fluid.

Describe the optimal carbohydrate replacement rate.

POST-EXERCISE CARBOHYDRATE INTAKE

Outline an optimal replenishment schedule of carbohydrate after prolonged, intense exercise.

List three foods that should be avoided when rapidly replenishing glycogen reserves because of their slow rates of intestinal absorption.

1.

2.

3.

CARBOHYDRATE NEEDS IN INTENSE TRAINING

Provide a nutrition-based explanation for "staleness" following repeated days of strenuous endurance training or competition.

DIET, GLYCOGEN STORES, AND ENDURANCE

In what way does preexercise diet effect performance?

A diet deficient in _____ rapidly depletes _____ and _____ glycogen.

CARBOHYDRATE LOADING

Briefly outline three stages for increasing muscle glycogen storage with the "classic" carbohydrate leading procedure.

Stage 1:

Stage 2:

Stage 3:

Limited Applicability

The benefits of carbohydrate loading apply to what type of physical activity?

Negative Aspects

Each gram of glycogen stored in liver and muscle results in an additional storage of ____ g of water.

Under what two conditions would carbohydrate loading be ill-advised?

1.

2.

Modified Loading Procedure

List two differences between the modified and classic carbohydrate loading procedure.

1.

2.

Practice Quiz

MULTIPLE CHOICE

1. Likely to burn the most total calories:
   a. Gymnastics practice
   b. Triathlon
   c. Daily swim practice for collegiate swimmers
   d. Marathon
   e. One mile run at maximal pace

2. Fruits and vegetables are rich sources of:
   a. Lipids
   b. Carbohydrates
   c. Proteins
   d. Lipoproteins
   e. b and d

3. At least how many hours are needed to restore muscle glycogen levels following prolonged exercise?
   a. 48
   b. 24
   c. 12
   d. 2
   e. None of the above

4. A high-protein, precompetition meal:
   a. Increases protein breakdown and subsequent fluid loss
   b. More difficult to digest than a high-carbohydrate meal
   c. Decreases total plasma cholesterol
   d. a and b
   e. a and c

5. Liquid meals are beneficial for all of the following reasons except:
   a. Easily digested
   b. Likely to provide sufficient energy without the feeling of fullness
   c. Contribute to fluid needs
   d. Promotes rapid and complete digestion
   e. Decrease blood lactate accumulation

6. The Eating Right Pyramid emphasizes:
   a. Grains and vegetables
   b. Fruits and meats
   c. Dairy products and grains
   d. Smaller portion sizes from plant sources
   e. Equal portion sizes from each food group

7. Ingesting low-glycemic carbohydrate in the immediate pre-exercise period may confer benefit by:
   a. Providing a source of "slow release" intestinal glucose during exercise
   b. Stimulating insulin release early in exercise
   c. Stimulating insulin release later in exercise
   d. Facilitating thermoregulation
   e. a and b

8. Prudent advice recommends ingesting _____ of water immediately before exercising in the heat:
   a. 4 to 6 L
   b. 400-600 mL
   c. 1000-2000 mL
   d. 2000-3000 mL
   e. 400-600 g

9. Concerning the precompetition meal:
   a. Avoid foods high in lipid and protein
   b. Avoid foods high in carbohydrate
   c. Include foods high in protein
   d. Include vitamin and mineral supplements
   e. Avoid fluids

10. Carbohydrates in either liquid or solid form consumed during exercise:
   a. Benefit performance in relatively high-intensity, long-term aerobic exercise and repetitive short bouts of near-maximal effort
   b. Do not benefit endurance performance
   c. Benefit only short-duration, sub-maximum effort
   d. Enhance aerobic exercise performance by increasing the $\dot{V}O_{2max}$
   e. a and d

TRUE / FALSE

1. _____ The glycemic index represents the blood glucose increases after ingesting a food containing 50 g of carbohydrate.

2. _____ Endurance athletes often experience "staleness" due to depletion of protein reserves.

3. _____ Fasting 24 hours prior to exercise provides benefits if the athlete "carbo-loads" the week before competition.

4. _____ Timing of the pre-event meal is important due to the body's limited protein reserves.

5. _____ It is wise to consume carbohydrate-rich foods as soon as possible after heavy exercise to speed the glycogen replenishment process.

6. _____ The ideal precompetition meal provides adequate carbohydrate energy and assumes optimal hydration.

7. _____ A "prudent diet" for a physically active person should contain about 60% of its calories in the form of high quality protein.

8. _____ One hour generally suffices to digest and absorb a meal.

9. _____ The ideal oral rehydration solution contains a carbohydrate concentration of between 5 and 8%.

10. _____ Drinking a strong sugar solution 30 minutes prior to exercise significantly improves endurance capacity.

CROSSWORD PUZZLE

CHAPTER 9

ACROSS

1. international endurance bicycle race
4. many physically active people actually _____ yet weight less
7. this type of meal is effective for pre-event feeding
8. reference values specifically for use on food labels; comprised of RDI and DRV values (abbr.)
9. famous living Greek marathoner
12. simple sugar that causes only a minimal insulin response with essentially no decline in blood glucose
13. system to classify food groups based on geometric shape
18. when derived from corn starch breakdown reduces osmolality and facilitates water movement from the stomach into the intestine for absorption
20. one approach to weight loss
21. minimum number of hours for glycogen replenishment
22. person weighing 77 kg requires this many grams of protein daily
23. glycogen synthesizing enzyme
24. concentration of particles in solution
25. exaggerated insulin release in response to sugar challenge

DOWN

2. fluid replacement should approach this percentage of sweat rate
3. replenish water loss
5. appraises the general "healthfulness" of one's diet (abbr.)
6. index of carbohydrate absorption
9. ultramarathoner
10. natives of Mexico who consume a diet high in complex carbohydrates
11. term for food consumed prior to competition
14. this intensity of aerobic exercise stresses glycogen reserves
15. eating right _____
16. daily standards for nutrients and food components (such as lipids and fiber) that are important to health; appear on food labels (abbr.)
17. animal polysaccharide
19. hormone secreted by beta cells of islets of Langerhans

The Pulmonary System and Exercise

1. 2,3-diphosphoglycerate

2. Acid

3. Acidosis

4. Alveolar air composition

5. Alveolar ventilation

6. Alveoli

7. Ambient air

8. Anatomical dead space

9. Arteriovenous oxygen difference

10. Base

11. Bicarbonate

12. Bohr effect

13. Bronchi

14. Bronchioles

15. Bronchodilators

16. Buffers

17. Carbamino compounds

18. Carbonic acid

19. Carbonic anhydrase

20. Chemoreceptors

21. Diaphragm

22. Dyspnea

23. Emphysema

24. Exercise-induced bronchospasm

25. Expiratory reserve volume

26. $FEV_{1.0}$ / FVC

27. Forced vital capacity

28. Functional residual capacity

29. Glottis

30. $H_2CO_3^-$

31. $[H+]$

32. Haldane effect

33. HCO_3^-

34. Hematocrit

35. Hemoglobin

36. Hyperpnea

37. Hyperventilation

38. Inspiratory muscles

39. Inspiratory reserve volume

40. Intrapulmonic pressure

41. Intrathoracic pressure

42. Maximum voluntary ventilation

43. Mechanoreceptors

44. Medulla

45. Minute ventilation

46. Myoglobin

47. OBLA

48. Oxyhemoglobin

49. Oxyhemoglobin dissociation curve

50. Partial pressure of gas

51. Percent saturation

52. Phase I ventilatory response

53. Phase II ventilatory response

54. Phase III ventilatory response

55. Physiologic dead space

56. P_{O_2}

57. Poiseoille's law

58. Pressure differential

59. Pulmonary ventilation

60. Residual lung volume

61. Sodium bicarbonate

62. Solubility

63. Spirometer

64. Tidal volume

65. Torr

66. Total lung capacity

67. Tracheal air

68. Ventilatory equivalent

69. Ventilatory system

70. Ventilatory threshold

PULMONARY STRUCTURE AND FUNCTION

List three functions of the ventilatory system.

1.

2.

3.

ANATOMY OF VENTILATION

Place the following terms in the order of air flow movement during the inspiratory cycle: <u>bronchioles</u>, <u>trachea</u>, <u>alveoli</u>, and <u>bronchi</u>.

1.   3.

2.   4.

Name the process whereby ambient air enters into and exchanges with the air in the lungs.

Lungs

Give the surface area dimension of the lungs.

Alveoli

What is the primary function of the alveoli?

Mechanics of Ventilation

What causes the changes in lung volume during inspiration and expiration?

Inspiration

List three muscles involved in inspiration.

1.

2.

3.

Expiration

List three factors that cause expiration of air from lungs.

1.

2.

3.

LUNG VOLUMES AND CAPACITIES

Static Lung Volumes

Draw and label a spirometer tracing to illustrate typical values for the various static lung volume measures. (Hint: Refer to Fig.10.5 in your textbook.)

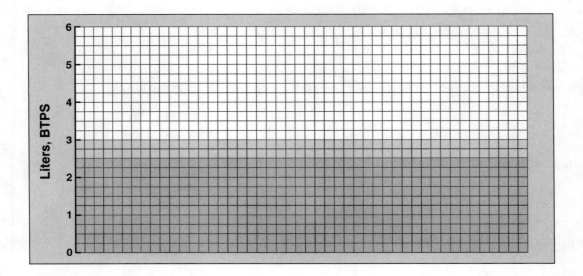

Residual Lung Volume

The RLV serves what important physiological function?

Dynamic Lung Volumes

List two factors that determine dynamic lung volumes.

1.

2.

Forced Expiratory Volume-to-Forced Vital Capacity Ratio.

Forced expiratory volume indicates what two components of lung function?

1.

2.

Maximum Voluntary Ventilation

Discuss whether the MVV exceeds the ventilation volume observed during maximal exercise.

PULMONARY VENTILATION

Minute Ventilation

Minute Ventilation = _____ × _____

Give the tidal volume increase during exercise as a percent of vital capacity.

Alveolar Ventilation

Calculate the alveolar ventilation per breath for a tidal volume of 600 mL.

What is the best way to increase alveolar minute ventilation, increasing depth, or rate of breathing? Explain.

Physiologic Dead Space

List two conditions that increase physiologic dead space.

1.

2.

Depth Versus Rate

Discuss the contributions of breathing rate and tidal volume to the increased ventilation during moderate and heavy exercise.

Moderate exercise

Heavy exercise

DISRUPTIONS IN NORMAL BREATHING PATTERNS

Dyspnea

What causes the sense of the inability to breathe during exercise, particularly in novice exercisers?

Hyperventilation

Describe the immediate physiologic effects of overbreathing (hyperventilation).

Valsalva Maneuver

Give the magnitude of intrathoracic pressure increase with a Valsalva maneuver?

Physiologic Consequences

List two physiologic consequences of a Valsalva maneuver.

1.

2.

GAS EXCHANGE

Give the percentage concentration of oxygen, carbondioxide and nitrogen in ambient air.

Partial pressure = _____ $\times$ _____

Ambient Air

Compute the partial pressures of O_2, CO_2, and N_2 of ambient air at sea level (barometric pressure = 760 mm Hg).

P_{O_2} = _____ $\times$ _____ = _____

P_{CO_2} = _____ $\times$ _____ = _____

P_{N_2} = _____ $\times$ _____ = _____

Tracheal Air

List the pressure exerted by water vapor and the inspired dry air molecules in tracheal air.

Water vapor

Dry air molecules

Alveolar Air

Give the average percent of O_2, CO_2 and N_2 in alveolar air at sea level.

$O_2 =$

$CO_2 =$

$N_2 =$

MOVEMENT OF GAS IN AIR AND FLUIDS

According to Henry's Law, the amount of gas that dissolves in fluid depends on what two factors?

1.

2.

Pressure

The pressure of a specific gas on both sides of a permeable membrane reaches equilibrium when:

_____

How do changes in gas partial pressure affect its movement in a liquid?

Solubility

Why is less pressure differential required to move a given amount of CO_2 into or out of a fluid compared with the same quantity of O_2?

GAS EXCHANGE IN THE BODY

What is the average partial pressure of O_2 and CO_2 in the alveoli at rest? How does heavy exercise affect these values? (Hint: Refer to Fig. 10.10 in your textbook.)

O_2:

CO_2:

Rest:

Heavy Exercise:

Gas Exchange in the Lungs

List three reasons that account for the dilution of oxygen in inspired air compared to ambient air.

1.

2.

3.

Physiologic Adjustments Limit Blood-Flow Velocity

What adjustments increase pulmonary blood flow volume with a proportionate increase in blood flow velocity?

Gas Exchange in Tissues

Give the average partial pressures of O_2 and CO_2 in tissues at rest. How does heavy exercise affect these values?

    PO_2:

    PCO_2:

    Rest:

    Heavy Exercise:

OXYGEN AND CARBON DIOXIDE TRANSPORT

OXYGEN TRANSPORT IN THE BLOOD

List two ways for oxygen transport in the blood.

1.

2.

Oxygen Transport in Physical Solution

Give the amount of oxygen carried in physical solution in:

    100 mL plasma:

    5000 mL plasma:

Oxygen Combined with Hemoglobin

Complete the formula for the oxygenation of hemoglobin to oxyhemoglobin:

$Hb + 4 O_2 \longrightarrow$

The oxygenation of hemoglobin to oxyhemoglobin depends upon what main factor?

Oxygen-Carrying Capacity of Hemoglobin

List the average hemoglobin values (g per 100 mL blood) for men and women.

Men:

Women:

Each gram of hemoglobin loosely combines with _____ mL of oxygen.

Po₂ and Hemoglobin Saturation

Draw and label the oxyhemoglobin dissociation curve. (Hint: Refer to Fig. 10.11 in your textbook.)

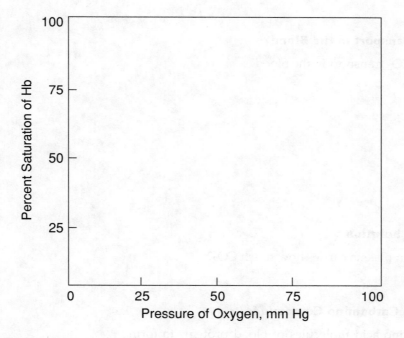

Po₂ in the Lungs

Hemoglobin is almost completely saturated with oxygen at approximately what Po₂?

Each 100 mL of blood leaving the lungs at sea level carries approximately how many mL of oxygen?

Tissue Po$_2$

In the tissues during rest, the blood's Po$_2$ falls to about _____ mm Hg, and the hemoglobin remains about _____ % saturated with oxygen.

Bohr Effect

List three factors that cause the oxyhemoglobin dissociation curve to shift downward and to the right.

1.

2.

3.

Myoglobin and Muscle Oxygen Storage

Complete the following equation for the oxygenation of myoglobin.

$$Mb + O_2 \longrightarrow$$

Carbon Dioxide Transport in the Blood

List three ways for CO$_2$ transport in the blood.

1.

2.

3.

Carbon Dioxide in Solution

Free CO$_2$ dissolved in plasma carries how much CO$_2$?

Carbon Dioxide as Carbamino Compounds

CO$_2$ reacts with amino acid molecules of blood proteins to form _____. The

globin portion of hemoglobin carries about _____ % of the body's total CO$_2$.

Carbon Dioxide as Bicarbonate

Complete the following equation to show the formation of bicarbonate in the tissues.

$$CO_2 + H_2O \longrightarrow$$

Plasma bicarbonate transports what percent of total CO$_2$ in the body?

REGULATION OF PULMONARY VENTILATION

VENTILATORY CONTROL

List two factors that regulate pulmonary ventilation at rest.

1.

2.

Neural Factors

In addition to the inherent activity of the respiratory center in the medulla, list three other neural mechanisms for ventilatory control at rest.

1.

2.

3.

Humoral Factors

The _____ regulates pulmonary ventilation at rest.

Plasma P_{O_2} and Chemoreceptors

Peripheral chemoreceptors sensitive to arterial blood's P_{O_2} are located in the _____

_____ and _____ .

Plasma P_{CO_2} and H^+ Concentration

Describe what happens to the resting pulmonary ventilation with an increase in inspired P_{CO_2}.

Hyperventilation and Breath-Holding

What primary factor causes the urge to breathe during breathholding?

VENTILATORY CONTROL IN EXERCISE

Chemical Control

Why can't chemical stimuli entirely explain hyperapnea during exercise?

Nonchemical Control

List three nonchemical factors to explain the rapid increase in ventilation at the onset of exercise.

1.

2.

3.

Neurogenic Factors

List two neurogenic factors that control pulmonary ventilation during exercise.

1.

2.

Influence of Temperature

What role does body temperature have in regulating pulmonary ventilation?

Integrated Regulation

Describe final phase Phase III of ventilatory control during exercise.

PULMONARY VENTILATION DURING EXERCISE

PULMONARY VENTILATION AND ENERGY DEMANDS

Ventilation in Steady-Rate Exercise

Draw and label the relationship between $\dot{V}O_2$ and pulmonary ventilation and blood lactate accumulation during graded exercise up to maximum levels. (Hint: Refer to Fig. 10.19 in your textbook.)

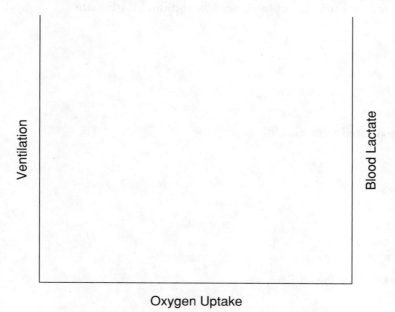

What term describes the ratio of minute ventilation to oxygen uptake? Present an average value for this measure for healthy young adults and children during submaximal exercise.

Term:

Average values:

Adults:

Children:

Ventilation in Non-Steady-Rate Exercise

Give a typical value for the ventilatory equivalent during intense exercise in healthy people.

Ventilatory Threshold

Describe the ventilatory threshold.

Onset of Blood Lactate Accumulation

At what average percent of $\dot{V}O_{2max}$ does blood lactate begin to accumulate in a healthy, untrained person during graded exercise?

Write the chemical reaction for the buffering of lactic acid by sodium bicarbonate.

List two chemical factors associated with the OBLA.

1.

2.

CAUSES OF OBLA

Give the most likely explanation for the OBLA.

OBLA AND ENDURANCE PERFORMANCE

In addition to $\dot{V}O_{2max}$, what other variable provides a consistent and powerful predictor of aerobic endurance performance?

What are the likely causes for the increase in OBLA with endurance training?

DOES VENTILATION LIMIT AEROBIC CAPACITY?

If pulmonary ventilation limits aerobic power, would an increase or decrease occur in the $\dot{V}_E/\dot{V}O_2$ ratio during graded exercise? Discuss.

Work of Breathing

List two factors that determine the energy requirement of breathing.

1.

2.

What percent of the total $\dot{V}O_2$ does the cost of breathing account for during rest and maximal exercise?

Rest:

Maximal exercise:

Effects of Cigarette Smoking

Describe the acute effects of cigarette smoking on airway resistance to breathing. How might this adversely affect exercise capacity?

Practice Quiz

MULTIPLE CHOICE

1. A volume of air remaining in the lungs after a maximum expiration:
   a. Expiratory reserve volume
   b. Tidal volume
   c. Residual volume
   d. Inspiratory reserve volume
   e. None of the above
2. Valsalva maneuver causes:
   a. A reduced venous return
   b. A decrease in blood pressure
   c. Dizziness
   d. a and b
   e. All of the above

3. Static lung volumes:
   a. Can predict endurance exercise performance
   b. Cannot predict endurance exercise performance
   c. Are larger for the untrained
   d. Are correlated to blood cholesterol levels
   e. a and d
4. Lung volume measurements should be:
   a. Expressed per unit body density
   b. Based on age and body mass
   c. Based on age, sex, and stature
   d. Expressed relative to percent body fat
   e. None of the above

5. Transports the major quantity of carbon dioxide:
   a. Free CO_2 in dissolved blood
   b. Combined with water to form bicarbonate
   c. In combination with myoglobin
   d. In combination with hemoglobin
   e. Carboxyglobin

6. _____ augments the blood's oxygen carrying capacity more than 60 times:
   a. Carbon dioxide
   b. Hemoglobin
   c. Plasma
   d. Bicarbonate
   e. Ferritin

7. Small amount of oxygen dissolved in plasma:
   a. Establishes oxygen partial pressure in blood, and thus determines oxygenation and deoxygenation of hemoglobin
   b. Increases blood acidity
   c. Determines alveolar air Po_2
   d. Ultimately combines with serum ferritin
   e. Provides a good measure of nitrogen elimination

8. Does not reduce the effectiveness of Hb to hold oxygen?
   a. Increased acidity
   b. Increased temperature
   c. Increased carbon dioxide concentration
   d. Increased plasma thyroxine
   e. None of the above

9. Pulmonary ventilation:
   a. Increases lineraly with $\dot{V}O_2$
   b. Decreases lineraly with $\dot{V}O_2$
   c. Is not related to $\dot{V}O_2$
   d. Relates more to HR than $\dot{V}O_2$
   e. C and D

10. OBLA
   a. Predicts heart rate during endurance exercise
   b. Effectively predicts endurance performance
   c. Effectively predicts $\dot{V}CO_2$ production during anaerobic exercise
   d. Can be measured using breathing rate
   e. None of the above

TRUE / FALSE

1. _____ Expiration moves air into the lungs.

2. _____ The anatomic dead space represents the portion of air involved in gaseous exchange with the blood.

3. _____ Alveolar ventilation in heavy exercise increases proportionately with oxygen uptake only if FVC is above average.

4. _____ Exercise training in the asthmatic can increase air flow reserve and reduce ventilatory work.

5. _____ Hemoglobin saturation changes little until the PO_2 falls below 60 mm Hg.

6. _____ Carbon dioxide is 25 times more soluble in plasma than oxygen.

7. _____ Carbamino compounds transport the major quantity of carbon dioxide in the body.

8. _____ For men and women, the point of OBLA represents a consistent and powerful predictor of aerobic exercise performance.

9. _____ Only one day of abstinence from smoking produces a substantial reversibility in the increased oxygen cost of breathing.

10. _____ Hyperpnea refers to a decreased minute ventilation.

CROSSWORD PUZZLE

CHAPTER 10

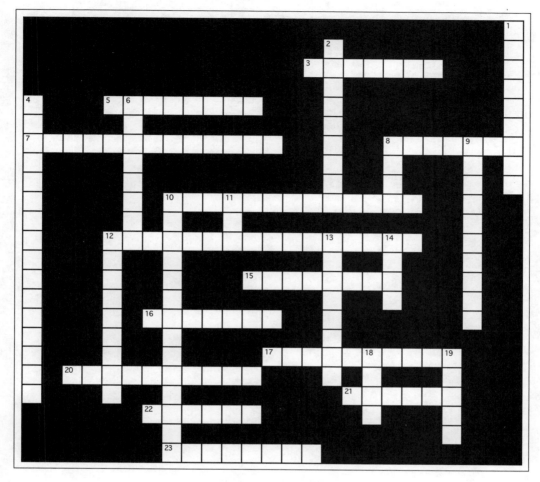

ACROSS

3. the specificity of the ventilatory response to training is related to a lower blood level of blood _____

5. term relating to base
7. most important respiratory stimulus
8. chemicals that minimize changes in [H+]
10. peripheral arterial chemoreceptors
12. overbreathing
15. _____ buffers provide the rapid first line of defense
16. respiratory center
17. minute ventilation + VO2 is termed the ventilatory _____
20. arterial PCO2 of about 50 mmHg is the _____ for breathhold
21. the point for OBLA is somewhat _____ than for the lactate threshold
22. brain area responsible for immediate, rapid rise in exercise ventilation
23. danish chemist who devised pH scale

DOWN

1. condition of abnormally low H+
2. compounds formed when CO2 combines with protein
4. peripheral receptors sensitive to muscle action
6. in light to moderate exercise ventilation increases _____ with VO2 and VCO2
8. binds with H+
9. with COPD, resistance is greatest in the _____ phase of the ventilatory cycle
10. sensitive to decreased arterial PO2 and PCO2
11. this many days of smoking abstinence reverses increased cost of breathing
12. another word for increased exercise ventilation
13. condition of abnormally high [H+]
14. onset of blood lactic acid buildup (abbr.)
18. liberates H+ in solution
19. phase of ventilatory control involving "fine tuning" through peripheral feedback mechanisms

The Cardiovascular System and Exercise

1. a-$\bar{v}$ O_2 difference

2. Acetylcholine

3. Adrenergic fibers

4. Afterload

5. Angina pectoris

6. Anticipatory heart rate

7. Arrhythmia

8. Arteries

9. Arterioles

10. Athlete's heart

11. Atrioventricular (AV) node

12. Auscultatory method

13. Baroreceptors

14. Bicuspid valve

15. Bradycardia

16. Capillaries

17. Cardiac output

18. Carotid artery

19. Catecholamines

20. Cholinergic fibers

21. CO_2 rebreathing

22. Concentric hypertrophy

23. Coronary circulation

24. CPR

25. Diastole

26. Diastolic blood pressure

27. Dye dilution

28. Eccentric hypertrophy

29. ECG

30. Echocardiography

31. Epinephrine

32. Extrasystole

33. Extrinsic regulation

34. Fick method

35. Frank-Starling mechanism

36. Hypertension

37. Max $\dot{Q}$

38. Myocardial infarction

39. Myocardium

40. Norepinephrine

41. Peripheral resistance

42. Poiseuille's law

43. Precapillary sphincter

44. Preload

45. Pulmonary circulation

46. Purkinje system

47. PVC

48. Rate-pressure product

49. S-A node

50. Semilunar valves

51. Sphygmomanometer

52. Stroke volume

53. Systemic circulation

54. Systole

55. Tachycardia

56. Thrombus

57. Tricuspid valve

58. Vagus nerve

59. Varicose veins

60. Vasomotor tone

61. Veins

62. Venous pooling

63. Venous return

64. Venous tone

65. Ventricular fibrillation

66. Venules

Study Questions

THE CARDIOVASCULAR SYSTEM

COMPONENTS OF THE CARDIOVASCULAR SYSTEM

List four important functions of the circulatory system.

1.

2.

3.

4.

Heart

What type of muscle makes up the myocardium?

List two functions of the left and right sides of the heart.

Left heart   Right heart

1.

2.

Name two different heart valves and give their functions.

Name   Function

1.

2.

Arteries

What is the main function of the arteries and arterioles?

Explain why arterioles are referred to as "resistance vessels."

Capillaries

How long does it take a blood cell to pass through a typical capillary?

Veins

Venous Return

What do veins have that arteries do not have?

Name the body's largest vein?

Give an advantage of having the venous system under low pressure.

A Significant Blood Reservoir

The venous system contains how much of the resting total blood volume?

Varicose Veins

What causes varicose veins?

Venous Pooling

Describe the reasons for fainting in the tilt table experiment?

ACTIVE COOL-DOWN

Why is it always beneficial to actively cool down after exercise compared to lying down?

BLOOD PRESSURE

Rest

Identify three locations to feel the pulse wave as blood passes through the arterial system?

1.

2.

3.

Give a normal systolic and diastolic blood pressure during rest.

During Exercise

Describe a normal systolic and diastolic blood pressure response during rhythmic aerobic exercise.

Rhythmic Exercise

Typically, how much does systolic and diastolic blood pressure increase during rhythmic muscular activities?

Resistance Exercise

Hypertensive individuals should not engage in heavy resistance training. Why?

Upper-Body Exercise

Describe the difference in blood pressure response during upper body exercise compared with walking.

Body Inversion

Should hypertensive individuals perform body inversion exercises? Why? Why not?

In Recovery

Why is aerobic exercise sometimes recommended for hypertensive individuals?

HEART'S BLOOD SUPPLY

What is unique about the heart's blood supply?

Myocardial Oxygen Utilization

How much oxygen is extracted from the blood in the coronary arteries at rest?

List two factors responsible for increasing coronary blood flow during intense exercise.

1.

2.

Rate-Pressure Product: An Estimate of Myocardial Work

Write the equation for the rate-pressure product (RPP).

The RPP relates highly to _____.

Heart's Energy Supply

What "anaerobic" macronutrient substrate can serve as energy fuel for the heart?

CARDIOVASCULAR REGULATION AND INTEGRATION
HEART RATE REGULATION

At what rate would the heart beat on its own without neural regulation?

Intrinsic Regulation

Where does the electrical impulse for the heart originate?

Identify the pacemaker of the heart.

Heart's Electrical Impulse

After originating at the S-A node, the electrical impulse spreads across the atria to a small knot of tis-

sue termed the _____.

Electrocardiogram

Draw and label a typical ECG tracing.

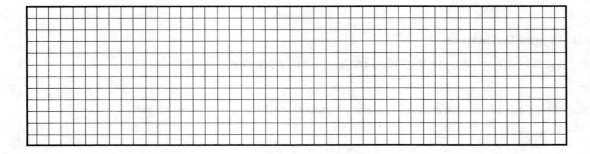

Extrinsic Regulation

Sympathetic Influence

Parasympathetic Influence

List two major neural influences on the heart.

1.

2.

What influence does the sympathetic system exert on heart rate?

What influence does the parasympathetic system exert on heart rate?

List two effects of exercise training on neural regulation of the heart.

1.

2.

Training Effects on Catecholamines

Explain why aerobic exercise training results in significant bradycardia.

Cortical Influence

Why does heart rate increase immediately prior to exercise?

PERIPHERAL INPUT

Describe the effect of an increase in blood pressure on heart rate?

CAROTID ARTERY PALPATION

Why isn't the carotid artery the best location to record heart rate under all conditions?

Arrhythmias

Heart Rate Irregularities

List three possible causes of extrasystoles.

1.

2.

3.

List three different types of heart rate irregularities.

1.

2.

3.

BLOOD DISTRIBUTION

Exercise Effects

What difference in renal blood flow exists between rest and maximum exercise.

Blood Flow Regulation

Blood volume is directly proportional to _____, and inversely related

to_____.

List the most important factor that affects blood flow.

Local Factors

Indicate two important functions served by opening dormant capillaries in exercise.

1.

2.

Neural Factors

How do adrenergic and cholinergic neural fibers effect the heart?

   Adrenergic fibers:

   Cholinergic fibers:

Hormonal Factors

List two hormones that cause a generalized systemic vascular constrictor response, except in the blood vessels of the heart and skeletal muscles.

1.

2.

Integrated Response in Exercise

Put a plus or minus to characterize the relative involvement of chemical, neural, and hormonal adjustments before and during exercise.

	Before exercise	During exercise
Chemical:		
Neural:		
Hormonal:		

CARDIOVASCULAR DYNAMICS DURING EXERCISE

CARDIAC OUTPUT

Write the equation for cardiac output.

Cardiac Output Measurement

Direct Fick Method

Write the equation for cardiac output using the Fick procedure.

Calculate the cardiac output for a person with a $\dot{V}O_2$ of 400 mL·min^{-1} and an a-$\bar{v}O_{2diff}$ of 7 mL O_2·100 mL^{-1}.

Other Less Invasive Methods

List two methods other than the Fick procedure to measure cardiac output.

1.

2.

RESTING CARDIAC OUTPUT

Untrained Persons

List a typical cardiac output for an <u>untrained</u> man and woman?

    Man:

    Woman:

Endurance Athletes

Does an endurance athlete have a higher or lower cardiac output compared to sedentary person? Discuss.

List a typical cardiac output for an endurance <u>trained</u> man and woman?

Man:

Woman:

EXERCISE CARDIAC OUTPUT

Draw and label a graph of cardiac output versus progressively increasing oxygen uptake during exercise.

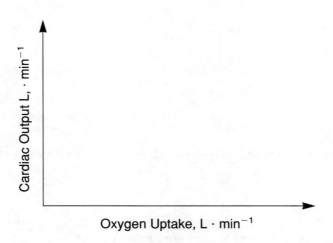

How can aerobically trained athletes achieve higher maximum cardiac output's than the untrained, despite a slightly reduced maximum heart rate?

EXERCISE STROKE VOLUME

Stroke Volume and $\dot{V}O_{2max}$

Draw and label the relationship between $\dot{V}O_{2max}$ and max stroke volume for individuals who vary widely in $\dot{V}O_{2max}$.

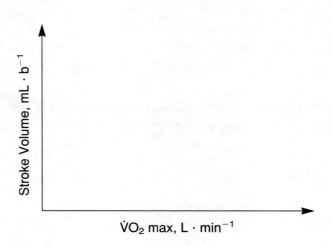

Stroke Volume: Increases

What two physiologic mechanisms regulate stroke volume of the heart?

1.

2.

GREATER SYSTOLIC EMPTYING VERSUS ENHANCED DIASTOLIC FILLING

What body position most likely enhances diastolic filling?

Complete the sentence "A greater systolic ejection, with or without an accompanying increase in end

diastolic volume, occurs because _____."

EXERCISE HEART RATE

Graded Exercise

Draw and label the relationship between heart rate and VO$_2$ during graded exercise for an untrained
and aerobically trained person.

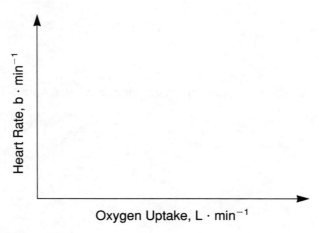

Submaximum Exercise

Cardiac Output Distribution

Rest Exercise

In the following table show the relative distribution (%) of the cardiac output during rest and exercise
to the skeletal muscles, digestive tract, liver, and kidneys.

	Rest	**Exercise**
Muscles		
Digestive tract		
Liver		
Kidneys		

Blood Flow Redistribution

Describe the redistribution of blood during exercise?

Blood Flow to the Heart and Brain

At rest, the myocardium normally uses _____% of the oxygen in the blood flowing through the coronary circulation.

CARDIAC OUTPUT AND OXYGEN TRANSPORT

How much oxygen is normally carried in each liter of arterial blood at sea level?

If 5 liters of blood circulates through the body each minute, how much total possible oxygen moves through the system?

Rest

How much oxygen returns unused to the heart during rest?

During Exercise

With a 16-L cardiac output during exercise, how much oxygen circulates each minute?

Maximum Cardiac Output and $\dot{V}O_{2max}$

Draw and label the relationship between $\dot{V}O_{2max}$ and maximum cardiac output.

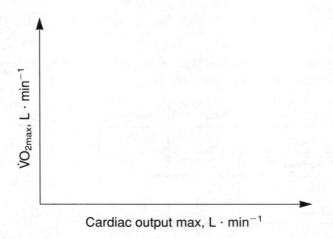

Gender Differences in Cardiac Output

Give one reason why females have a larger cardiac output at the same absolute submaximum $\dot{V}O_2$ compared to males?

EXTRACTION OF OXYGEN: THE a-$\bar{v}O_2$ DIFFERENCE

Write the equation for calculating the a-$\bar{v}O_2$ difference.

During Rest and Exercise

Give an average a-$\bar{v}O_2$ difference during rest and maximum exercise.

Rest:

Maximum Exercise:

At rest, how much of the oxygen returns "unused" to the heart in the mixed-venous blood?

Discuss how aerobic training increases the a-$\bar{v}O_2$ difference.

In Heart Disease

How can a heart disease patient improve aerobic capacity if there is little training improvement occurs in maximum cardiac output?

CARDIOVASCULAR ADJUSTMENTS TO UPPER-BODY EXERCISE

$\dot{V}O_{2max}$ attained during arm exercise equals about what percent of the value attained during leg exercise?

"ATHLETE'S HEART"

List two structural differences between the heart of a healthy untrained person and an the heart of an endurance athlete.

1.

2.

Discuss whether the "Athlete's Heart" syndrome is dangerous to overall health?

Practice Quiz

MULTIPLE CHOICE

1. At rest, the myocardium extracts approximately _____ % of the oxygen flowing through the coronary arteries.
   a. 30
   b. 60
   c. 80
   d. 100
   e. None of the above

2. Following exercise, blood pressure:
   a. Remains elevated for one to two hours
   b. Immediately returns to the pre-exercise level
   c. Remains elevated only for ten to fifteen minutes
   d. Falls below pre-exercise levels for several hours or more
   e. None of the above

3. The main energy substrates for the heart include all of the following except:
   a. Protein
   b. Fatty acids
   c. Glucose
   d. Lactic acid
   e. a and d

4. Movement of large quantities of blood from peripheral veins into the central circulation:
   a. Venous dilation
   b. Venous constriction
   c. Arterial stability
   d. Arterial constriction
   e. a and c

5. Cholinergic fibers release _____ and serve as _____.
   a. Norepinephrine, vasodilators
   b. Acetycholine, vasoconstrictors
   c. Acetycholine, vasodilators
   d. Norepinephrine, vasoconstrictors
   e. None of the above

6. Each 100 mL of arterial blood at sea level carries about _____ mL of oxygen.
   a. 10
   b. 20
   c. 50
   d. 65
   e. More than 65

7. Cardiac hypertrophy occurs in response to?
   a. Heart attack
   b. Sedentary lifestyle
   c. Increased frequency of exercise
   d. Increased work load placed on the myocardium
   e. All of the above

8. Which is true about resting cardiac output?
   a. About the same for endurance trained and untrained
   b. Greater for endurance trained than untrained
   c. Greater for untrained than endurance trained
   d. None of the above

9. The average a-$\bar{v}O_2$ difference between arterial blood and mixed-venous blood equals:
    a. 5 mL of $O_{2/10}$ mL of blood
    b. 5 mL of $O_{2/100}$ mL of blood
    c. 10 mL of $O_{2/50}$ mL of blood
    d. 30 mL of $O_{2/100}$ mL of blood
    e. None of the above

10. Other than the active muscle mass, which organ receives the greatest redistribution of the blood flow during moderate exercise?
    a. Kidney
    b. Brain
    c. Liver
    d. Skin
    e. None of the above

TRUE / FALSE

1. _____ Cardiac rhythm initiates at the A-V node.

2. _____ The refractory period serves an important function because it provides sufficient time for ventricular filling between heart beats.

3. _____ Adrenergic fibers stimulate vasoconstriction.

4. _____ Afterload represents a more complete emptying during systole despite increasing systolic pressure.

5. _____ Approximately 4-7 mL of blood each minute reaches every 100 grams of muscle at rest.

6. _____ The apparent gender difference in cardiac output occurs because men typically have lower heart rates than women.

7. _____ The product of heart rate and diastolic blood pressure provides a good estimate of myocardial workload

8. _____ During isometric, free weight, and hydraulic exercise, peak systolic and diastolic pressures mirror the hypertensive state.

9. _____ Systolic blood pressures are greater for work performed with the legs than with the arms.

10. _____ The inferior vena cava empties blood into the lower half of the body.

CROSSWORD PUZZLE

CHAPTER 11

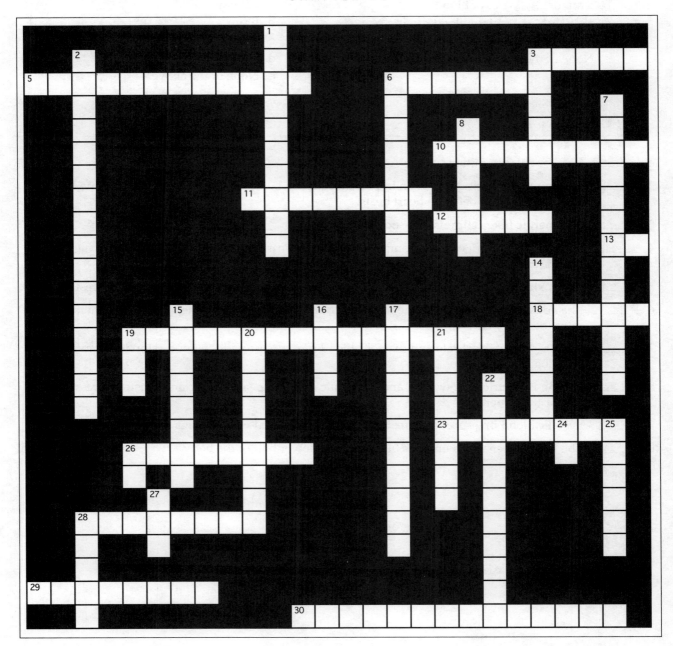

ACROSS

3. Endurance training increases the influence of this nerve on the heart
4. nerve that slows heart rate
5. high blood pressure
6. contraction phase of cardiac cycle
9. small arterial vessel that is increased with aerobic training
10. blood vessel where diffusion occurs

11. valve prevents backflow from left ventricle into left atrium
12. atypical accumulation of fluid in the interstitial space
13. Volume of blood pumped by left ventricle in one minute (abbr.)
18. major artery from the heart
19. ultrasound technique to study heart structure
23. deficiency of blood tissue supply

26. diversion of blood from one body region to another
28. diseased vein
29. veins leading into the right atrium
30. chemicals that act as neurotransmitters (dopamine, epinephrine, and norepinephrine)

DOWN

1. another word for heart muscle
2. used to measure blood pressure
3. small vein
6. one of the men who described force-length characteristics of cardiac muscle
7. receptors sensitive to changes in blood pressure
8. pacemaker
14. relaxation phase of cardiac cycle

15. blood clot that forms in the vascular circuit
16. man who developed invasive method for measuring cardiac output
17. slow heart rate
19. graphic record of electrical changes during cardiac cycle
20. small artery supplies blood to capillaries
21. specialized fibers that speed impulse over ventricles
22. rapid heart rate
24. heart attack (abbr.)
25. carries blood away from the heart
26. amount of blood pumped by the left ventricle each beat (abbr.)
27. bundle of _____
28. blood vessels that contain valves

The Neuromuscular System and Exercise

Define Key Terms and Concepts

1. Actin

2. Actin-myosin orientation

3. Action potential

4. Actomyosin

5. Afferent neuron

6. All-or-none principle

7. Alpha motoneurons

8. Anterior motoneuron

9. Ascending nerve tracts

10. Autonomic nervous system

11. Autonomic reflex arc

12. Axon

13. Baroreceptors

14. Brain stem

15. Central nervous system

16. Cerebellum

17. Chemoreceptors

18. Cholinesterase

19. Cross bridges

20. Cross-extensor reflex

21. Dendrites

22. Diencephalon

23. Dorsal horn

24. Efferent neurons

25. Endomysium

26. Enervation ratio

27. Epimysium

28. EPSP

29. Excitation-contraction coupling

30. Extrafusal fibers

31. Extrapyramidal tract

32. Fast-twitch muscle fiber

33. Fibrils/myofibrils

34. Filaments/myofilaments

35. FOG fibers

36. Gamma efferent fibers

37. Golgi tendon organs

38. Hypothalamus

39. Intracellular tubule system

40. Intrafusal fibers

41. IPSP

42. Limbic system

43. Motor unit

44. Motor unit recruitment

45. Muscle spindles

46. Myelin sheath

47. Myofibrils

48. Myofilaments

49. Myosin

50. Myosin ATPase

51. Neurilemma

52. Neuromuscular fatigue

53. Neuromuscular junction

54. Neuropeptides

55. Nitric oxide

56. Nodes of Ranvier

57. Pacinian corpuscles

58. Parasympathetic nervous system

59. Perimysium

60. Periosteum

61. Peripheral nervous system

62. Propreoceptors

63. Pyramidal tract

64. Pyramidal tract

65. Reflex arc

66. Reflex inhibition

67. Reticular formation

68. Rigor mortis

69. Sarcolemma

70. Sarcomere

71. Sarcoplasm

72. Sarcoplasmic reticulum

73. Schwann cells

74. Size principle

75. Sliding-filament theory

76. Slow-twitch muscle fiber

77. Somatic nerves

78. Spinal cord

79. Stretch reflex

80. Striated

81. Sympathetic nervous system

82. Telencephalon

83. Temporal summation

84. Tendons

85. Transverse tubule system

86. Triad

87. Tropomyosin

88. Troponin

89. T-tubule system

90. Type I fibers

91. Type II fiber

92. Type IIa fibers

93. Type IIb fibers

94. Type IIc fiber

95. Ventral horn

NEURAL CONTROL OF HUMAN MOVEMENT

NEUROMOTOR SYSTEM ORGANIZATION

List two major parts of the human nervous system.

1.

2.

Central Nervous System – The Brain

List seven main areas of the brain and give one fact about each.

Main Area   Fact

1.

2.

3.

4.

5.

6.

7.

Organization of the Brain

Give one fact about each of the four major anatomic divisions of the brain.

                                    Fact

Telencephalon (Cerebrum):

Diencephalon:

Mesencephalon (Midbrain):

Metencephalon:

LIMBIC SYSTEM

List the two activities that the limbic system likely influences.

1.

2.

Central Nervous System – The Spinal Cord

There are _____ vertebrae in the human spinal column.

Give the functional role of the ventral and dorsal horns of the spinal cord.

   Ventral horn:

   Dorsal horn:

Ascending Nerve Tracts

The ascending nerve tracts send _____ information from _____

receptors to the _____.

SENSORY RECEPTORS

List four factors monitored by the body's peripheral sensory receptors.

1.   3.

2.   4.

Descending Nerve Tracts

Name and give the function of the two major descending nerve tract pathways.

Pathway   Function

1.

2.

List four structures that interconnect with the reticular formation.

1.   3.

2.   4.

PYRAMIDAL TRACT

The pyramidal tract neurons excite what neurons?

EXTRAPYRAMIDAL TRACT

List two functions of the extrapyramidal neurons.

1.

2.

Brain Neurotransmitters

List four important brain neurotransmitters and give one fact about each.

Neurotransmitter   Fact

1.

2.

3.

4.

Peripheral Nervous System

List two types of peripheral efferent nerves.

1.

2.

What always results from somatic efferent nerve firing?

List the number of pairs of spinal nerves for the following regions of the spine cord.

Cervical: Sacral:

Thoracic: Coccygeal:

Lumbar:

The peripheral nervous system consists of ___ pairs of spinal nerves.

Somatic Nervous System

Somatic nerves enervate _____ muscle.

Somatic nerves always cause _____ to activate muscle.

Autonomic Nervous System

Autonomic nerves always activate _____ muscle.

Describe the major function of the autonomic nervous system.

SYMPATHETIC AND PARASYMPATHETIC NERVOUS SYSTEMS

Indicate the areas of the body innervated by the sympathetic and parasympathetic nervous system.

Sympathetic:

Parasympathetic:

Autonomic Reflex Arc

Draw and label a typical autonomic reflex arc in the spinal cord. (Hint: Refer to Fig. 12.5 in your text-book.)

Complex Reflexes

List the first two events in the crossed-extensor reflex.

1.

2.

Learned Reflexes

Cite one difference between automatic reflexes and learned reflexes.

Nerve Supply to Muscle

Do nerves usually innervate "one" or "one or more" muscle fibers?

Motor Unit Anatomy

List two components of a motor unit.

1.

2.

Anterior Motoneuron

List and give the function for three parts of an anterior motoneuron.

Part   Function

1.

2.

3.

Explain how the structure and location of motor units produce a more effective application of force and diminished mechanical stress.

Neuromuscular Junction (Motor End-Plate)

Indicate the primary function of the neuromuscular junction.

Excitation

What role does acetylcholine play in neuromuscular excitation?

Facilitation

Describe situations in which temporal summation could enhance exercise performance.

Inhibition

What happens if a motoneuron is subjected to both excitatory and inhibitory influences with a correspondingly large IPSP?

MOTOR UNIT PHYSIOLOGY

Fill-in the following table concerning the physiologic and mechanical properties of muscle fiber types and motor units.

Motor Unit Type	Force Production	Contraction Speed	Fatigue Resistance	Muscle Fiber Type
Fast Fatigable (FF)				
Fast-Fatigue Resistant (FR)				
Slow (S)				

Twitch Characteristics

List three categories of motor units with respect to speed, force, and fatigue characteristics.

1.

2.

3.

Tension-Generating Characteristics

All-or-None Principle

State the all-or-none principle in 20 words or less.

Gradation of Force

Discuss two factors contributing to ability to vary force of muscular contraction.

1.

2.

Motor Unit Recruitment

Describe the size principle of motor unit recruitment.

Neuromuscular Fatigue

List four factors that could explain muscular fatigue during exercise.

1.   3.

2.   4.

PROPRIOCEPTORS IN MUSCLES, JOINTS, AND TENDONS

Proprioceptors monitor these three variables:

1.

2.

3.

Muscle Spindles

Describe the major function of muscle spindles.

How may types of specialized fibers does the muscle spindle contain?

Stretch Reflex

List three main components of the stretch reflex.

1.

2.

3.

Golgi Tendon Organs

Describe the major function of Golgi tendon organs.

The Golgi tendon organ is activated by _____ and/or _____.

Pacinian Corpuscles

The Pacinian corpuscles respond to _____ and/or _____.

MUSCULAR SYSTEM: ORGANIZATION AND ACTIVATION
COMPARISON OF SKELETAL, CARDIAC, AND SMOOTH MUSCLE

List two structural and functional characteristic of skeletal, cardiac, and smooth muscle.

Type	Structural Characteristic	Functional Characteristics
Skeletal	1. 2.	1. 2.
Cardiac	1. 2.	1. 2.
Smooth	1. 2.	1. 2.

GROSS STRUCTURE OF SKELETAL MUSCLE

Draw and label the cross section of a typical muscle. (Hint: Refer to Fig. 12.14 in your textbook.)

Chemical Composition

List the percentage composition of skeletal muscle for:

Water:

Protein:

Other:

Blood Supply

There are approximately _____ capillaries per mm^2 of skeletal muscle.

Describe the effect of straining-type muscular activities on blood flow in active muscle.

Muscle Capillarization

Capillarization of skeletal muscle is about _____% greater in endurance athletes than in the un-trained.

SKELETAL MUSCLE ULTRASTRUCTURE

Draw a diagram showing the microscopic anatomy of a muscle sarcomere. Identify the actin and myosin filaments, the I and A bands, Z line, and H zone. (Hint: Refer to Figs. 12.5 and 12.16 in your textbook.)

The Sarcomere

What structure within the sarcomere provides the mechanical mechanism for muscle action?

Actin-Myosin Orientation

A single muscle fiber contains approximately how many thick and thin filaments?

What term best describes the projections of various proteins that form the contractile filaments?

Intracellular Tubule Systems

Describe the functional role of the tubule system network within a muscle fiber.

CHEMICAL AND MECHANICAL EVENTS DURING CONTRACTION AND RELAXATION

Sliding-Filament Theory

List five important facts in the sequencing of muscle contraction according to the sliding-filament theory.

1.   4.

2.   5.

3.

Mechanical Action of Crossbridges

What part of the myosin provides the mechanical power stroke for actin and myosin?

Link Between Actin, Myosin, and ATP

Complete the following equation:

Actomyosin + ATP $\longrightarrow$

Excitation-Contraction Coupling

Describe calcium's role in excitation-contraction coupling.

Relaxation

Describe the dynamics of intracellular calcium when a muscle is no longer stimulated.

Sequence of Events in Muscle Excitation-Contraction

Describe the eight main events in muscular contraction and relaxation in five words or less for each event.

1.   5.

2.   6.

3.   7.

4.   8.

MUSCLE FIBER TYPE

List two muscle fiber types and classify each by its contractile and metabolic characteristics.

   Type   Contractile Characteristics   Metabolic Characteristics

1.

2.

Measurement of Fiber Types

List two methods to classify muscle fiber types.

1.

2.

Fast-Twitch Fibers

List two types of physical activities that activate fast-twitch fibers?

Fast-Twitch Subdivisions

List three subdivisions of fast-twitch muscle fibers.

1.

2.

3.

Slow-Twitch Fibers

List two types of physical activities that activate slow-twitch muscle fibers?

Differences Between Athletic Groups

List the reported fiber-type distribution for the following groups of elite athletes. (Hint: Refer to Figs. 12.23 and 12.24 in your textbook.)

Athlete Group   **Fiber-Type Distribution**

Endurance runners:

Sprinters:

High jumpers:

Weight lifters:

Cross-country skiers:

Practice Quiz

MULTIPLE CHOICE

1. The pyramidal tract regulates:
   a. Postural movements
   b. Background level of neuromuscular tone
   c. Discrete musculature movements
   d. Facial movements
   e. None of the above

2. Basic neural component for controlling automatic muscular movements:
   a. Cerebellum
   b. Basal ganglia
   c. Reflex arc
   d. Motor unit
   e. None of the above

3. Determines number of muscle fibers in a motor unit:
   a. Muscle size
   b. Muscle strength
   c. Muscle's movement function
   d. Muscle fiber-to-neuron ratio
   e. c and d

4. Not commonly associated with fast-twitch muscle fibers:
   a. Large motoneurons with fast conduction velocities
   b. High peak tension
   c. Fatigue-resistance
   d. Great force
   e. None of the above

5. Pacinian corpuscles:
   a. Detect quick movement and deep pressure
   b. Detect differences in muscle tension
   c. Detect changes in muscle fiber length
   d. Detect motor unit firing patterns
   e. a and d

6. Muscle cell functional unit:
   a. Motor unit
   b. Sarcomere
   c. Endomysium
   d. Neuron
   e. None of the above

7. The sliding-filament theory proposes that muscles shorten or lengthen because:
   a. Thick and thin myofilaments slide past each other and change length
   b. Thick and thin myofilaments slide into each other
   c. Thick and thin myofilaments slide past each other and retain length
   d. Thick and thin myofilaments expand and contract as they enter the sarcomere
   e. None of the above

8. Which statements are true:
   a. Type I fibers are fast-twitch
   b. Type IIa fibers are fast-twitch
   c. Type IIb fibers are slow-twitch
   d. Type IIc fibers are the most common
   e. b and d

9. Fiber type with high capacity for aerobic and anaerobic energy transfer:
   a. FG
   b. SO
   c. FOG
   d. FO
   e. FGT

10. Which is true about fiber type and exercise training:
    a. Fiber type can change from slow to fast-twitch
    b. Fiber type can change from fast to slow-twitch
    c. Fiber type can change from type I to type II
    d. Fiber type does not alter to any large degree
    e. None of the above

T R U E / F A L S E

1. _____ The anterior motoneuron transmits the electrochemical nerve impulse from the spinal cord to the muscle.

2. _____ The IPSP excites the neuron, increasing its likelihood of firing.

3. _____ Alterations in muscle fiber type explain a large portion of the strength improvements with resistance training.

4. _____ Interneurons receive information from the reflex arc's sensory root.

5. _____ The reticular formation transmits impulses that inhibit neurons to the antigravity muscles involved in postural control.

6. _____ Protein makes up 75% of skeletal muscle.

7. _____ An increase in capillary density of aerobically trained muscles contribute to improved endurance capacity.

8. _____ Tropomyosin has a high affinity for calcium ions.

9. _____ Endurance activities activate fast-twitch muscle fibers.

10. _____ Exercise training does not improve the metabolic capacity of a specific muscle fiber type.

C R O S S W O R D   P U Z Z L E

ACROSS

1. _____ pattern of motor unit firing occurs in weight lifting
3. blood flow occluded when muscle generates about this percent of its force generating capacity
5. _____ twitch fibers have a high level of myosin ATPase
6. functional unit of muscle fiber
8. muscle bone attachment distal to body
11. tracing of muscle's electrical activity
14. basic unit of involuntary neural control
18. law that states that a motor unit either responds completely or not at all to a stimulus
20. physiological cross sectional area (abbr.)
21. this effect on neurons makes them harder to fire (abbr.)
23. the Nodes of _____ speed the transmission of the nerve impulse
30. a large efferent neuron that innervates striated muscle
37. the reflex _____ is comprised of a sensory neuron, spinal interneurons, and anterior motoneuron
39. pertaining to the body
40. intermediate type of fast-twitch fiber with high oxidative capacity (abbr.)
41. repeating pattern of two vesicles and T-tubules in region of each Z-line
42. component of nervous system associated with forebrain; concerned with emotional behavior and learning

44. tracing of brain's electrical activity (abbr.)
45. connects both ends of the muscle to the outermost covering of the skeleton
46. autonomic nervous system (abbr.)
47. fast-twitch muscle fibers
48. receptors stimulated by chemical state of fluid bathing them

DOWN

2. nervous system that includes brain and spinal cord (abbr.)
4. slow-twitch muscle fibers
5. a spindle-shaped type of muscle with expanded belly
7. part of the brain responsible for posture, locomotion, and equilibrium
8. neurons that relay information to various levels of the cord
9. a nerve cell
10. muscle form resembling a feather
12. first word of sensory receptors in ligaments
13. this effect on neurons makes them easier to fire (abbr.)
15. crucial mineral in muscle
16. transmits nerve impulses away from the cell body of neuron
17. lipid-protein sheath around axon
18. these motorneurons control skeletal muscle activity
19. high _____ is produced in FF motor units

CHAPTER 12

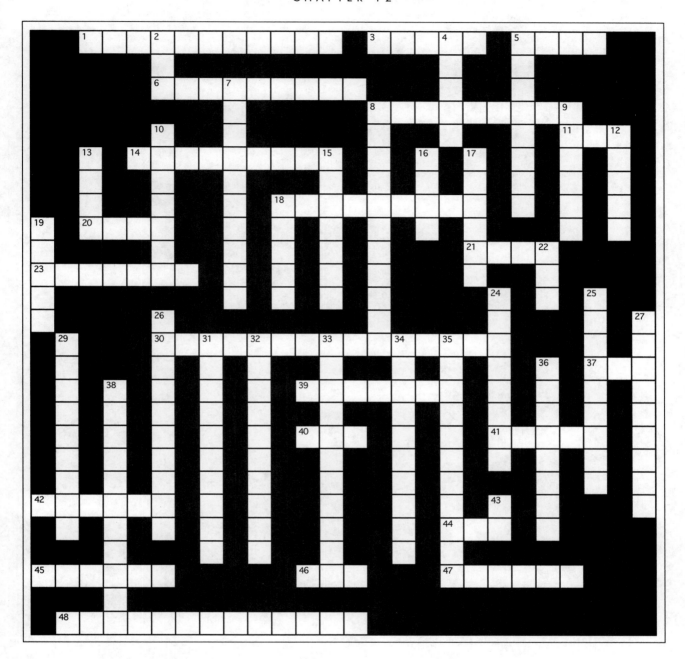

22. nervous system that includes spinal nerves (abbr.)
24. transmits nerve impulses towards the cell body of neuron
25. nerve tract that transmits impulses from brain to spinal cord
26. fiber type with high capability for power activities
27. bundle of up to 150 muscle fibers
28. functional unit of muscle
29. the anterior motoneuron and specific muscle fibers it innervates
31. layer of connective tissue surrounding fasciculus

32. interaction of this protein filament causes movement
33. muscle protein of actin filaments
34. specialized fibers attached in parallel to regular muscle fibers
35. adding more motor units to increase force production is called motor unit _____
36. skeletal muscle is this classification
38. globular "lollipoplike" head on myosin filaments
43. Muscle fiber type with fast contraction speed and high anaerobic power (abbr.)

Chapter 13

Hormones, Exercise, and Training

1. Acini

2. Acromegly

3. ACTH

4. Adenylcylase

5. ADH

6. Adrenal cortex

7. Adrenocortical hormones

8. Aldosterone

9. Allosteric modulation

10. Androgens

11. Angiotensin

12. Adrenal medulla

13. Beta cells

14. Beta-endorphin

15. Catecholamines

16. Cortisol

17. Cyclic AMP

18. Diabetes mellitus

19. Down-regulation

20. Endocrine gland

21. Erythropoietin

22. Exocrine gland

23. FPG

24. FSH

25. Glucagon

26. Glucocorticoids

27. HGH

28. Hormonal stimulation

29. Hormone

30. Host gland

31. Humoral stimulation

32. Insulin

33. Insulin antagonist hormone

34. Insulin resistance

35. Islets of Langerhans

36. LH

37. Mineralocorticoids

38. Neurohypophysis

39. NIDDM

40. Opioid peptides

41. Oxytocin

42. Parathyroid glad

43. Pituitary gland

44. PRL

45. Prostaglandins

46. PTH

47. Renin

48. Renin-angiotensin mechanism

49. Somatotropin

50. Insulin antagonist

51. Thyroxin

52. Thyroxin (T4)

53. Triiodothyronine (T3)

54. TSH

55. Type 1 diabetes

56. Type 2 diabetes

57. Up-regulation

58. Vasopressin (ADH)

Study Questions

ENDOCRINE SYSTEM OVERVIEW

Endocrine System Organization

List three components that characterize the endocrine system.

1.

2.

3.

What distinguishes a gland as endocrine or exocrine?

Nature of Hormones

List the three chemical categories of hormones.

1.

2.

3.

How Hormones Function

Indicate the major function of hormones.

1.

2.

3.

What determines a target cell's ability to respond to a hormone?

Hormone Effects on Enzymes

List three ways that hormones increase enzyme activity.

1.

2.

3.

Control of Hormone Secretion

List three ways endocrine glands become stimulated and give one fact about each.

Stimulated by   Fact

1.

2.

3.

RESTING AND EXERCISE-INDUCED ENDOCRINE SECRETIONS

ANTERIOR PITUITARY HORMONES

List six different hormones secreted from the pituitary gland.

1.   4.

2.   5.

3.   6.

Growth Hormone

List the major function of hGH.

Explain how hGH contributes to one's ability to perform endurance exercise and respond to training.

   Endurance exercise:

   Training response:

Exercise, hGH, and Tissue Synthesis

Explain how hGH responds to exercise to increase protein synthesis (and subsequent muscle hypertrophy).

Thyrotropin

Describe the major function of TSH.

Corticotropin

Describe the major function of ACTH.

Gonadotropic Hormones

List two gonadotropic hormones and their major function.

Hormone   Function

1.

2.

Prolactin

Describe the major function of PRL?

Endorphins

Endorphins are considered an _____ neurotransmitter.

POSTERIOR PITUITARY HORMONES

List two posterior pituitary hormones and their major function.

Hormone   Function

1.

2.

THYROID HORMONES

List two thyroid hormones and their major function.

Hormone   Function

1.

2.

PARATHYROID HORMONE

State the main function of parathyroid hormone.

ADRENAL HORMONES

List the adrenal gland's two structures.

1.

2.

Adrenal Medulla Hormones

List two adrenal medulla hormones and their major function.

Hormone   Function

1.

2.

Adrenal Cortex Hormones

List three groups of adrenocortical hormones and their major function.

Hormone   Function

1.

2.

3.

Mineralocorticoids

Name the most physiologically important mineralocorticoid.

Glucocorticoids

Describe two important functions of cortisol.

1.

2.

Androgens

Name the male and female sex organ and one hormone they produce.

	Sex organ	Hormone
Male:		
Female:		

PANCREATIC HORMONES

Name two different pancreatic hormone-producing tissues.

1.

2.

Insulin

Describe the major function of insulin in carbohydrate regulation and lipid metabolism.

Carbohydrate regulation:

Lipid metabolism:

Glucose-Insulin Interaction

Identify the substance that controls insulin secretion.

Describe the effects of the catecholamines on insulin release.

Glucagon

Why is glucagon known as the "insulin antagonist hormone?"

Blood Glucose Homeostasis and Hormone Action

List three events that decrease circulating blood glucose (Hint: Refer to Fig. 13.12 in your textbook.)

1.

2.

3.

List three events that increase circulating blood glucose (Hint: Refer to Fig. 13.13 in your textbook.)

1.

2.

3.

DIABETES MELLITUS

List and describe two major classifications of diabetes mellitus.

1.

2.

Tests for Diabetes

Describe the most common test for diabetes mellitus.

What level of fasting blood glucose suggests diabetes mellitus?

Type 1 Diabetes

List two characteristics of type 1 diabetes.

1.

2.

Type 2 Diabetes

List two characteristics of type 2 diabetes.

1.

2.

List three risk factors for type 2 diabetes.

1.

2.

3.

Diabetes and Exercise

List three exercise guidelines for type 1 diabetics.

1.

2.

3.

Exercise Guidelines for Type 1 Diabetes

List 5 exercise guidelines that type 1 diabetics should be aware of:

1.   4.

2.   5.

3.

Exercise Benefits for Type 1 Diabetics

List 5 benefits of exercise for type 1 diabetics.

1.   4.

2.   5.

3.

Exercise Benefits For Type 2 Diabetes

List three benefits of regular exercise for type 2 diabetics.

1.

2.

3.

Exercise Risks for Diabetics

List three potential adverse effects of exercise for the diabetic patient.

1.

2.

3.

ENDURANCE TRAINING AND ENDOCRINE FUNCTION

Discuss whether endurance training generally produces an increase or decrease in hormonal response to standard exercise.

List six different hormones and the response of each to exercise training. (Hint: Refer to Table 13.2 in your textbook.)

Hormone   Training Response

1.

2.

3.

4.

5.

6.

Anterior Pituitary Hormones

hGH and Long-Term Exercise Training

Describe the effect of long-term exercise on resting 24-hour hGH production.

Corticotropin ACTH

What effect does long-term exercise training have on ACTH levels?

PRL

What effect does long term exercise training have on PRL levels?

FSH, LH, and Testosterone

List two factors that contribute to menstrual dysfunction with exercise training, in addition to alterations in FSH and LH.

1.

2.

Posterior Pituitary Hormones

Vasopressin ADH

What benefit would a decreased ADH response have to regular exercise training?

Oxytocin

List one function of oxytocin.

Thyroid Hormones

Discuss whether the increased turnover of thyroid hormones with endurance training posess a long-term health risk.

Adrenal Hormones

Aldosterone

Describe the effect of exercise training on aldosterone resting levels.

Cortisol

Describe the effect of exercise training on cortisol resting levels.

Epinephrine and Norepinephrine

Describe the typical sympathoadrenal training response.

Pancreatic Hormones

Insulin and Glucagon

List two ways that regular aerobic exercise blunts the insulin response during rest and moderate exercise.

1.

2.

Exercise Training in Diabetes

Does regular aerobic exercise provide greater benefits for type 1 or type 2 diabetics?

RESISTANCE TRAINING AND ENDOCRINE FUNCTION

List two hormones involved in resistance training adaptations and indicate possible sex differences in hormone release.

Hormones   Sex Difference

1.

2.

Special Endocrine Considerations

List three opioid peptide hormones.

1.

2.

3.

Opioid Peptides and Exercise

Explain the so called "exercise euphoria" from moderate to intense exercise.

Practice Quiz

MULTIPLE CHOICE

1. Exocrine glands:
   a. Have no ducts
   b. Secrete their substances directly into the extracellular spaces around a gland
   c. Pancreas, an example
   d. Under nervous system control
   e. Excrete electrolytes

2. Which is not a method of hormone action for altering rates of specific cellular reactions of "target cells?"
   a. Decreasing rate of intracellular proteins synthesis
   b. Increasing enzyme activity rate
   c. Modifying cell membrane transport
   d. Inducing secretory activity
   e. a and b

3. Provides a stimulus to endocrine glands:
   a. Changing blood pH levels
   b. Hormonal stimulation
   c. Changing plasma ion and nutrient levels
   d. Neural stimulation
   e. All of the above

4. Which is false about growth hormone?
   a. Promotes cell division and cellular proliferation throughout the body
   b. Facilitates protein synthesis
   c. Secretion decreases with increasing exercise intensity
   d. Secretion decreases carbohydrate utilization
   e. All of the above

5. Major function of ADH:
   a. Stimulates ovaries to secrete estrogens
   b. Controls water excretion by kidneys
   c. Stimulates contraction of uterus muscle
   d. Stimulates carbohydrate and fat metabolism
   e. None of the above

6. Which is not a function of cortisol?
   a. Stimulates protein breakdown to amino acid in all cells except liver
   b. Controls sodium reabsorption in kidneys
   c. Supports action of other hormones in gluconeogenesis
   d. Inhibits glucose uptake and oxidation
   e. c and d

7. Which is false about insulin?
   a. Regulates glucose metabolism in all tissues except the brain
   b. Directly controlled by glucose level of blood passing through the pancreas
   c. Causes protein synthesis in cells
   d. Mobilizes fatty acids for use in place of sugar

8. Benefit of regular exercise for type 2 diabetes mellitus:
   a. Improved glycemic control
   b. Improved cardiovascular function
   c. Weight loss
   d. Disease prevention
   e. All of the above

9. Which is true about the pancreatic hormones and regular aerobic exercise:
   a. Decreases insulin sensitivity
   b. Lowers insulin requirement
   c. Causes larger insulin output to clear excess glucose
   d. Causes blood insulin and glucagon to increase above resting values
   e. None of the above

10. Exercise-induced elevation of this opioid associated with euphoria and increased pain tolerance:
    a. Beta-endorphins
    b. Beta-angiotensin
    c. Dyatropin
    d. a and c
    e. b and c

T R U E / F A L S E

1. _____ A target cell's ability to respond to a hormone depends largely on the presence of specific protein receptors.

2. _____ The adrenal gland secretes at least six different hormones and influences the secretion of several others.

3. _____ ACTH concentration increase with exercise duration if intensity exceeds 25% of aerobic capacity.

4. _____ Blood levels of "free" T4 increase during exercise.

5. _____ Aldosterone regulates calcium reabsorption in the kidneys distal tubules.

6. _____ Pancreatic alpha cells secrete insulin and pancreatic beta cells secrete glucagon.

7. _____ Type 1 diabetes usually occurs in younger individuals.

8. _____ Endurance training produces a decline in the magnitude of hormonal responses to a standard exercise load.

9. _____ Epinephrine and norepinephrine concentrations decrease during the first weeks of training.

10. _____ Testosterone and growth hormone represents two primary hormones involved in resistance training adaptations.

C R O S S W O R D   P U Z Z L E

C H A P T E R   1 3

ACROSS

4. these exocrine cells in the pancreas secrete digestive enzymes
6. islets composed of alpha and beta cells
8. initials for hormone that stimulates production of estrogen by ovaries
9. _____ cagon is the "insulin antagonist"
10. cells of pancrease that secrete insulin
11. produced by beta cells of pancreas
13. _____ erior portion of pituitary gland resembles true neural tissue
15. chemical messenger
17. inhibits testosterone (abbr.)
18. another word for pituitary gland
22. adrenocortical hormones that regulate electrolytes

26. generally occurs in overweight, sedentary middle-aged individuals with a family history of the disease
28. stimulates thyroxin (abbr.)
30. tissue classified as either endocrine or exocrine
32. _____ cells are cells on which a hormone acts
33. male sex hormones
34. adrenal glands and ovaries (in females) and testes (in males) produce the sex steroid hormones
36. renin stimulates release of this kidney hormone
37. another name for hGH
38. another name for growth hormone

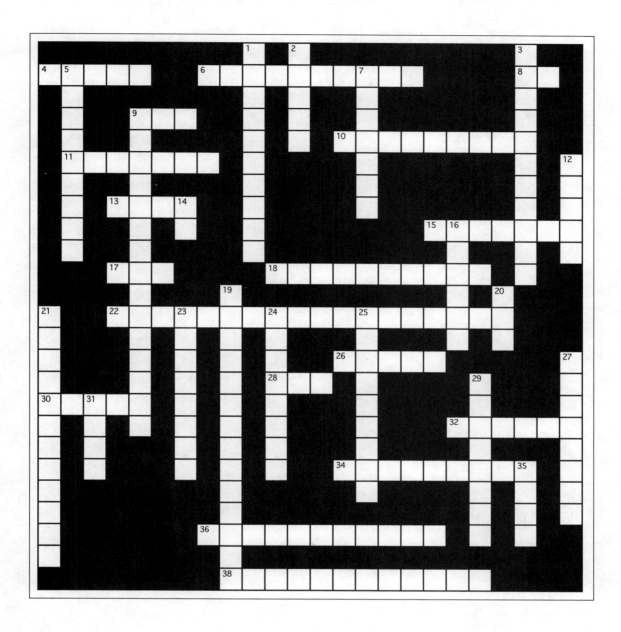

DOWN

1. beta _____ are opioid polypeptides
2. reduced renal blood flow stimulates the kidneys to release this hormone into the blood
3. promotes retention of sodium by kidneys
5. Second messenger to facilitate hormone function
7. cap like gland above each kidney
9. stimulate gluconeogenesis and serve as an insulin antagonist; produced by adrenal cortex
12. insulin deficiency due to destroyed pancreas' insulin-producing beta cells
14. Triiodothyronine (abbr.)
16. type of peptide that exerts morphine like effect

19. epinephrine and norepinephrine
20. a gonadotropic hormone
21. target cells form more receptors in response to increasing hormone levels
23. type of gland such as sweat gland
24. hormone in birthing and lactation
25. T4
27. underproduction causes of this hormone Addison's disease
29. eisease caused by inadequate insulin
31. stimulates production and release of cortisol
35. testosterone is a _____ oid type hormone

Training the Anaerobic and Aerobic Energy Systems

Define Key Terms and Concepts

1. a-$\bar{v}O_2$ difference

2. Aerobic energy system

3. Age-predicted HR_{max}

4. Anaerobic energy system

5. Continuous training

6. Conversational exercise

7. Criterion-referenced standards

8. Eccentric hypertrophy

9. Fartleck training

10. Fetal blood supply

11. Four factors affecting aerobic conditioning

17. Individual difference principle

12. Glycolytic capacity

18. Interval training

13. Health-related fitness

19. Karvonen method

14. Heart rate reserve

20. Lactate stacking

15. High energy phosphates

21. Lactate threshold training

16. HR threshold

22. LSD training

23. Norm-referenced standards

24. Overload principle

25. Overtraining syndrome

26. Rating of perceived exertion (RPE)

27. Recommended training frequency

28. Relative exercise stress

29. Relief interval

30. Reversibility principle

31. SAID principle

32. Specificity principle

33. Staleness

34. Stroke volume

35. Tecumseh step test

37. Ventilatory equivalent

36. Training-sensitive zone

Study Questions

TRAINING MUST FOCUS ON ENERGY REQUIREMENTS

List three sports requiring high levels of aerobic energy transfer and three sports requiring high levels of anaerobic energy transfer.

Aerobic	*Anaerobic*
1.	
2.	
3.	

Energy for Exercise: Knowing What To Train For

Cite two exercise examples that support each of the three energy systems:

Energy System	*Exercise Example*
ATP-PC System	1.
	2.
Glycolytic System	1.
	2.

Aerobic System   1.

                                                  2.

The relative contribution to the total energy requirement differs markedly depending on exercise

_____ and _____.

GENERAL TRAINING PRINCIPLES

Identify and give an illustration of each of the four principles of training.

   Principle   *Illustration*

1.

2.

3.

4.

Overload Principle

List four variables manipulated to achieve an exercise overload.

1.   3.

2.   4.

Specificity Principle

Give an example to support the statement: "Specific exercise elicits specific adaptations creating specific training effects."

Specificity of Aerobic Power: An Example

Give one example of exercise-training specificity for aerobic performance.

Local Adaptations

Explain how aerobic overload of specific muscle groups contribute to enhanced exercise performance and aerobic power in endurance training.

Individual Differences Principle

Give one example why the principle of individual differences is important in formulating an exercise program.

Reversibility Principle

What are the effects of detraining on $\dot{V}O_{2max}$? (Hint: Refer to Figure 14–4 in your textbook.)

ANAEROBIC TRAINING

List three physiologic changes that occur with sprint- and power-type training.

1.

2.

3.

ATP-PCr System

Give an example of training the ATP-PCr system for a football defensive back.

Glycolytic (Lactic Acid) System

Define "lactate stacking" and describe its importance for training the glycolytic system?

AEROBIC TRAINING

Evaluating Initial Status and Training Success

Discuss any difference in training activities for a person desiring sport-specific improvements versus improved general fitness.

The Gold Standard: $\dot{V}O_{2max}$

Briefly explain differences between "norm-referenced" and "criterion-referenced" standards for $\dot{V}O_{2max}$.

Cardiorespiratory Fitness Standards for Children

Explain why run-walk performance change in children does not adequately reflect changes in $\dot{V}O_{2max}$.

Heart Rate Response

Draw and label a curve showing heart rate response during submaximal exercise and recovery for an aerobically trained athlete and sedentary individual. (Hint: Refer to Fig. 14.5 in your textbook.)

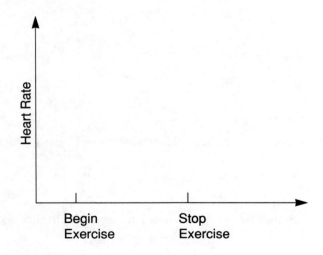

List one important point when using heart rate to establish training intensity for upper-compared to lower-body exercise.

Responders and Nonresponders

Explain the large variations in $\dot{V}O_{2max}$ improvement between people undergoing similar training?

Exercise Adherence

List three factors related to good adherence success to regular exercise.

1.

2.

3.

Tecumseh Step Test for Cardiorespiratory Fitness

Outline the procedure for administering the Techumseh Step Test.

FACTORS THAT AFFECT AEROBIC CONDITIONING

List four factors that influence the response to aerobic conditioning.

1.   3.

2.   4.

Initial Level of Cardiovascular Fitness

How much can $\dot{V}O_{2max}$ be expected to improve during a 12-week aerobic training program?

Frequency of Training

What is generally considered the minimum frequency per week necessary to induce an aerobic training effect?

Duration of Training

Give the generally recommended minimum threshold of exercise duration necessary to induce an aerobic training effect.

Intensity of Training

List five ways to express exercise intensity.

1.   4.

2.   5.

3.

What is meant by the term "training sensitive zone?"

Overly Strenuous Exercise Not Necessary

What is generally considered the "ceiling" training intensity for percent HR_{max} and percent $\dot{V}O_{2max}$.

% Heart rate

$\%\dot{V}O_{2max}$

How Long Before Improvement Occurs?

Discuss the general time course for improvement in aerobic capacity with training. (Hint: Refer to Fig. 14.10 in your textbook.)

ADAPTATIONS TO EXERCISE TRAINING

List what you consider the three most important metabolic and physiologic adaptations to regular aerobic training.

1.

2.

3.

Anaerobic System Changes

List three physiologic changes that occur with sprint- and power-type training.

1.

2.

3.

Aerobic System Changes

Discuss whether training-induced changes in the aerobic system occur independently of sex and age.

Metabolic Adaptations

METABOLIC MACHINERY

List two changes in skeletal muscle that improves its ability to generate ATP by oxidative phosphorylation.

1.

2.

ENZYMES

Give a possible explanation why changes in enzymes with training enhances endurance performance.

FAT METABOLISM

Give two reasons for increased lipolysis with aerobic training.

1.

2.

CARBOHYDRATE METABOLISM

List two reasons for enhanced carbohydrate breakdown with aerobic training.

1.

2.

MUSCLE FIBER TYPE AND SIZE

Which muscle fiber type shows the greatest hypertrophy with aerobic training?

Cardiovascular Adaptations

HEART SIZE AND PLASMA VOLUME

List one objective change in heart size and plasma volume resulting from aerobic training.

Heart Size

Plasma Volume

STROKE VOLUME

List three differences in stroke volume response between aerobically trained and untrained individuals.

1.

2.

3.

HEART RATE

Graph the relationship between heart rate and oxygen uptake for trained and untrained individuals throughout the major portion of the exercise range. (Hint: Refer to Fig. 14.3 in your textbook).

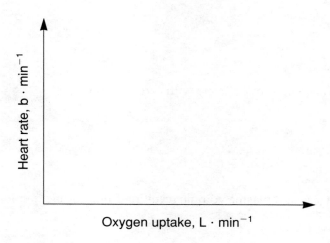

CARDIAC OUTPUT

Give the physiologic reason for increased cardiac output capacity resulting from aerobic training.

OXYGEN EXTRACTION

List two reasons for an increased a-$\bar{v}O_2$ differences resulting from aerobic training.

1.

2.

BLOOD FLOW AND DISTRIBUTION

List three reasons why aerobic training substantially increases muscle blood flow during maximal exercise.

1.

2.

3.

BLOOD PRESSURE

Give the expected change in resting systolic and diastolic blood pressure with regular aerobic exercise.

Systolic

Diastolic

Give one reason why aerobic training lowers resting blood pressure.

Pulmonary Adaptations

MAXIMAL EXERCISE AND SUBMAXIMAL EXERCISE

Describe changes in minute ventilation during submaximum and maximum exercise resulting from aerobic training.

Submaximal exercise

Maximal exercise

Describe the change in ventilatory equivalent during submaximal exercise from aerobic training.

SPECIFICITY OF VENTILATORY RESPONSE

Discuss the change that upper body aerobic training would cause in the ventilatory equivalent during submaximal treadmill walking.

Other Adaptations

BODY COMPOSITION

List the major body composition change resulting from aerobic training.

BODY HEAT TRANSFER

List the major reason well-hydrated, trained individuals exercise more comfortably in hot environments.

FORMULATING AN AEROBIC TRAINING PROGRAM
Guidelines

List three general guidelines when initiating an aerobic training program.

1.

2.

3.

Guidelines for Children

List one exercise training guideline used with children.

Establishing Training Intensity

Method 1. Train at a Percentage of $\dot{V}O_{2max}$

Give an example showing how to establish training intensity using percent $\dot{V}O_{2max}$.

Method 2. Train at Percentage of Maximum Heart Rate

Calculate your percent $\dot{V}O_{2max}$ training level from your age, and your ability to train at 80% of your HR_{max}. (Hint: refer to Fig. 14.16 in your textbook.)

ADJUST FOR SWIMMING AND OTHER UPPER BODY EXERCISES

How many $b \cdot min^{-1}$ must be subtracted from age-predicted HR_{max} when calculating heart rate training levels for swimming and other upper body exercises.

Swimming

Upper body exercise

Effectiveness of Less Intense Exercise

What would be a prudent training % HR_{max} for a middle-age person who starts an aerobic training program?

Train at a Perception of Effort

Describe the procedure to determine training intensity based on how you "feel" during exercise?

Train at the Lactate Threshold

Discuss the effect of exercising at the lactate threshold on the aerobic training response.

CONTINUOUS VERSUS INTERMITTENT AEROBIC TRAINING

Continuous Training

Define "continuous training" and indicate the group best suited for this training method.

Definition:

Group:

Interval Training

What four factors formulate an interval training exercise prescription?

1.   3.

2.   4.

Outline the method to determining the exercise "interval" in interval training.

Outline the method for determining the relief interval in interval training.

Rationale for Interval Training

Describe a practical method for determining the appropriate exercise and relief intervals for interval training. (Hint: Refer to Table 14.5.)

Formulating the Exercise: Relief Interval

List one guideline to formulating the exercise: relief interval.

Exercise:

Relief:

MAINTAINING AEROBIC FITNESS

List the most important factor for maintaining aerobic capacity.

EXERCISE TRAINING DURING PREGNANCY

Energy Cost and Physiologic Demands of Exercise

How much "extra" strain if any, does pregnancy add to the exercise effort?

Fetal Blood Supply

Does 30 to 40 min of moderate aerobic exercise compromise fetal blood flow?

Give prudent recommendations for duration, intensity and exercise mode during pregnancy.

Duration:

Intensity:

Mode:

Pregnancy Course and Outcome

List four outcome variables that are significantly greater in the off-spring of women who exercise throughout pregnancy compared to offsprings from sedentary mothers.

1.   3.

2.   4.

Practice Quiz

MULTIPLE CHOICE

1. Major objective of exercise training?
   a. Maintain ideal body weight
   b. Protect against cardiovascular diseases
   c. Cause biologic adaptations to physiologic systems to improve functioning
   d. Increase endurance
   e. None of the above

2. Maximum heart rate:
   a. Lower in arm exercise
   b. Higher in arm exercise
   c. Equal for arm and leg exercise
   d. Higher in older than younger individuals
   e. Higher in women than men

3. Maximum heart rate during swimming averages ___ $b \cdot min^{-1}$ lower than running.
   a. 2–5
   b. 6–9
   c. 10–13
   d. 14–17
   e. Greater than 17

4. Which is *true* about regular aerobic training's effect on blood pressure?
   a. Reduces systolic, increases diastolic blood
   b. Reduces diastolic, increases systolic
   c. Reduces systolic and diastolic
   d. Increases systolic and diastolic
   e. None of the above

5. Which is *not* a training method to improve aerobic fitness?
   a. Fartleck
   b. Interval
   c. Resistance
   d. Continuous

6. Which is *not* a metabolic change with short duration power-type training?
   a. Increase in $\dot{V}O_{2max}$
   b. Increase in resting levels of anaerobic substrates
   c. Increase in the quantity and activity of anaerobic enzymes
   d. Increase in the capacity to generate high blood lactate levels

7. Conditioning the aerobic systems occurs as long as exercise heart rate falls within ___% of HR_{max}.
   a. 10–30
   b. 30–50
   c. 50–70
   d. 70–90
   e. 90–100

8. Which is *not* an adaptation to aerobic training?
   a. Mitochondria increase ATP generating capacity
   b. Increase in reliance on fat metabolism
   c. Increase in level of aerobic system enzymes
   d. Trained muscle exhibits a greater capacity to oxidize carbohydrate
   e. Large increase in fat-free body mass

9. Which is *true* about cardiovascular adaptations to aerobic exercise?
   a. Cardiac hypertrophy occurs in the right ventricule
   b. Plasma volume increases within 4 to 5 training sessions
   c. Heart's stroke volume decreases during rest and during exercise
   d. Decreases in maximum cardiac output
   e. All of the above

10. Which is *false* about pulmonary adaptations to aerobic exercise training?
    a. Increase the amount of oxygen extracted from arterial blood
    b. Decrease in a-$\bar{v}$ O_2 difference
    c. Increase in $\dot{V}O_{2max}$
    d. Decrease pulmonary ventilation during submaximal exercise.
    e. All of the above are true

TRUE/FALSE

1. _____ Overload, specificity, individual differences, and reversibility describe the four training principles.

2. _____ Decreased capacity for blood lactate accumulation occurs following all-out exercise training.

3. _____ Aerobic training produces large increases in total muscle blood flow during maximal exercise.

4. _____ Components of the overload principle include frequency, intensity duration, and body composition.

5. _____ Aerobically trained muscle exhibits a greater capability to oxidize fats than untrained muscle.

6. _____ Specific exercise elicits general adaptations which create general training effects.

7. _____ The beneficial effects of exercise training are transient and reversible.

8. _____ The step test provides a useful method for evaluating the efficiency of the cardiovascular response to aerobic exercise

9. _____ Moderate aerobic exercise compromises fetal oxygen supply even for the physically active, healthy pregnant woman.

10. _____ A minimum of 90 minutes per exercise session is desirable for aerobic exercise training.

CROSSWORD PUZZLE

ACROSS

2. term for VO2max
3. about 70% to 90% of max HR is called the training _____
7. lower exercise intensity is offset by longer exercise _____
10. continuous activities of 3 min and longer engage mainly this system of energy transfer
14. this muscle thickens with pressure overload
16. acid produced during repetitive bouts of near maximum exercise
17. anaerobic training requires a high level of this psychological factor
19. all-out exercise for 60 s engages mainly this energy pathway
23. a rest: _____ interval is a period of passive recovery in interval training
24. this principle operative when training stops
25. lower recovery heart rate on this weight bearing exercise test indicates a higher aerobic fitness

DOWN

1. step test named for this New York College
2. max heart rate decreases with this variable
3. principle related to specificity of training (abbr.)
4. 70% max HR is the minimum threshold for the training sensitive _____
5. blood will do this in periphery tissues if exercise stops abruptly
6. method for estimating training heart rate that uses "heart rate reserve"
8. first word in abbreviation RPE

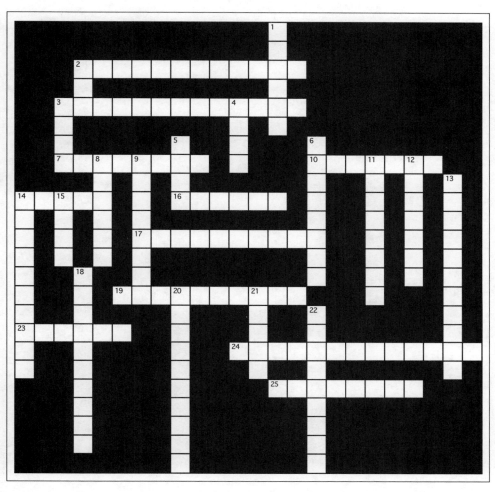

CHAPTER 14

9. this step test is named for a city in Michigan
11. a fundamental training principle
12. the lower the _____ fitness level, the greater the potential for improvement
13. endurance swimmers are not necessarily good endurance runners because this training principle
14. repetitive stair sprinting will train the _____ phosphates
15. initials for sports medicine organization
18. people respond differently to training. This is the _____ differences principle
20. type of training used by endurance athletes
21. recommended minimum number of days per week for aerobic training
22. second word of abbreviation RPE

Training Muscles to become Stronger

Define Key Terms and Concepts

1. Absolute muscular strength

2. Accommodating resistance exercise

3. Adaptation

4. Bidirectional muscle exercise

5. Cable tensiometry

6. Circuit resistance training (CRT)

7. Concentric action

8. Connective tissue damage

9. Delayed-onset muscle soreness (DOMS)

10. Disinhibition

11. Drop-jumping

12. Dynamic constant external resistance (DCER)

13. Dynamometry

14. Eccentric muscle action

15. EMG

16. Force-velocity relationship

17. Hyperplasia

18. Hypertrophy

19. Isokinetic muscle action

20. Isometric muscle action

21. Isotonic muscle action

22. Longitudinal splitting

23. Low back pain syndrome

24. Macrocycle

25. Miometric

26. Muscle cross-sectional area (MCSA)

27. Muscle hyperplasia

28. Muscular hypertrophy

29. Muscular overload

30. One-repetition maximum

31. Overload principle

32. Periodization

33. Plyometrics

34. Power strength training zone

35. Progressive resistance exercise (PRE)

36. Psyching

37. Relative strength

38. ROM

39. Spasm hypothesis

40. Sticking point

41. Strength training specificity

42. Strength training zone

43. Stretch-shortening cycle

44. Tear theory

45. Type I resistance equipment

46. Type II resistance equipment.

47. Type III resistance equipment

48. Voluntary maximal muscle action

Study Questions

MUSCULAR STRENGTH: MEASUREMENT AND IMPROVEMENT

FOUNDATIONS FOR STUDYING MUSCULAR STRENGTH

List four areas for which the study of muscular strength development provides practical applications.

1.

2.

3.

4.

Objectives of Resistance Training

List three objectives of resistance training for overall fitness and exercise performance.

1.

2.

3.

MEASUREMENT OF MUSCULAR STRENGTH

List four methods to measure muscular strength.

1.

2.

3.

4.

Cable Tensiometry

List two advantages of cable tensiometry testing.

1.

2.

Dynamometry

Describe the principle of operation for the hand-grip and back-lift steel dynamometers.

Hand-grip:

Back-lift:

One-Repetition Maximum

Describe the procedure for determining the 1-RM.

Computer-Assisted Electromechanical and Isokinetic Determinations

Describe a major advantage of isokinetic strength testing.

STRENGTH TESTING CONSIDERATIONS

List four factors affecting strength testing results regardless of assessment method.

1.   3.

2.   4.

Physical Testing in the Occupational Setting

Outline the ideal method for assesing muscular strength in the occupational setting.

Resistance Training Equipment

List the three categories of resistance training equipment.

1. Category I

2. Category II

   a.

   b.

3. Category III

TRAINING MUSCLES TO BECOME STRONGER

Overload and Intensity

How is the amount of overload usually expressed in resistance training?

List three approaches to applying overload in resistance training.

1.

2.

3.

What minimum overload intensity induces strength gains?

Muscle Actions

List and describe the three types of muscle action.

Action   Description
1.

2.

3.

What is a more useful term than "isotonic" to describe resistance training in which external resistance remains constant?

Force:Velocity Relationship

Draw a graph showing the force-velocity relationship for concentric and eccentric actions. (Hint: refer to Fig. 15.4 in your textbook.)

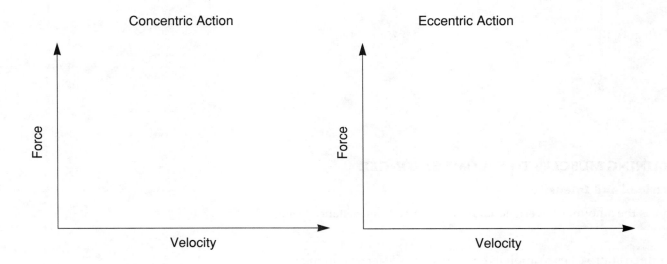

Power:Velocity Relationship

Draw and label a graph showing the relationship between power and velocity. (Hint: refer to Fig. 15.5 in your textbook.)

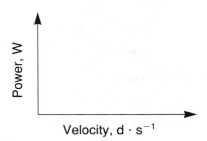

Force:Power-Range of Motion Relationship

Draw a graph showing the relationship between force:power versus range of joint motion for arm curls. (Hint: refer to Fig. 15.6 in your textbook.)

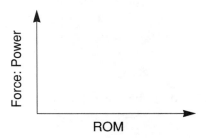

Load:Repetition Relationship

Draw a graph showing the relationship between maximum number of repetitions and resistance load at 20 to 100% 1-RM. Indicate the strength training sensitive zone. (Hint: refer to Fig. 15.7 in your textbook.)

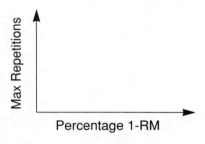

GENDER DIFFERENCES IN MUSCULAR STRENGTH

Absolute Strength

What are the gender differences in absolute muscular strength of the upper and lower body?

Upper:

Lower:

Relative Strength

Quantify the maximum muscle force (N) generated by human skeletal muscle per square cm of muscle cross-sectional area.

List three ways to express relative strength.

1.

2.

3.

RESISTANCE TRAINING FOR CHILDREN

List three considerations before initiating a children's resistance training program.

1.

2.

3.

SYSTEMS OF RESISTANCE TRAINING

List five different systems for muscular strength development.

1.   4.

2.   5.

3.

Isometric Training (Static Exercise)

List four facts regarding isometric strength training.

1.   3.

2.   4.

Limitations of Isometrics

List two limitations of isometric strength training.

1.

2.

Benefits of Isometrics

For what purpose is isometric exercise best suited?

Dynamic Constant External Resistance (DCER) Training

Progressive Resistance Exercise

Briefly describe the principles of the progressive resistance exercise system.

Variations of PRE

List five general observations regarding optimal number of sets and repetitions, and frequency and relative intensity required to improve muscular strength.

1.   4.

2.   5.

3.

Responses of Men and Women to DCER Training

What average percentage improvement in strength can be expected for DCER training?

Variable Resistance Training

List three factors why a single variable resistance "cam" does not allow for individual differences in mechanics and force applications.

1.

2.

3.

Special Consideration: The Lower Back

What role can resistance training play in reducing the risk of low back pain syndrome?

Isokinetic Training

Explain the unique aspects of isokinetic resistance training compared with more "standard" forms of resistance training.

Experiments With Isokinetic Exercise and Training

Which type of muscle fibers produce the greatest isokinetic force?

Fast- Versus Slow-Speed Isokinetic Training

Which improves "general" strength more, fast- or slow-speed isokinetic training?

Plyometric Training

Give the general rationale for plyometric training.

Practical Applications

Describe an example of plyometric training for a track sprint athlete.

Research With Plyometric Exercises

What type of performance benefits most from plyometric training?

Injury and Other Considerations

Can plyometric training cause greater injury compared with other modes of resistance training?

Comparison of Training Systems

Comparisons of strength training systems generally support the _____ and _____ principles of strength training.

Practical Implications of Specificity

Give one example of strength training specificity.

Periodization

Describe the purpose of fractionating the resistance training cycle in periodization.

Should periodization be sport specific? Discuss.

ADAPTATIONS TO RESISTANCE TRAINING

List five factors that impact development and maintenance of muscle mass.

1.   4.

2.   5.

3.

NEURAL ADAPTATIONS

List three neurobehavioral factors that undergo adaptation with regular resistance training.

1.

2.

3.

List three possible factors for enhanced neural adaptations with resistance training.

1.

2.

3.

Motor Unit Activation: Size Principle

List two factors that produce a continuum of voluntary force output from muscle.

1.

2.

List the conditions that type I and type II motor units become activated?

Type I:

Type II:

MUSCLE ADAPTATIONS

What is the most visible adaptation to resistance training?

Muscle Fiber Hypertrophy

List four muscle adaptations that help to explain muscle growth form resistance training.

1.   3.

2.   4.

Muscle Remodeling: Can Fiber Type Be Changed?

State the current consensus regarding changes in muscle fiber type with resistance training.

Under what condition could muscle fiber remodeling occur?

Muscle Hypertrophy and Testosterone Levels

Indicate the current state of research regarding testosterone and muscular strength development.

Muscle Hypertrophy: Males Versus Female

Discuss whether the skeletal muscle of women can hypertrophy to the same extent as men with regular resistance training.

Muscle Fiber Hyperplasia

Summarize the current state of knowledge regarding muscle fiber hyperplasia with resistance training.

CONNECTIVE TISSUE AND BONE ADAPTATIONS

Summarize the current state of knowledge regarding changes in connective tissue and bone with resistance training.

CARDIOVASCULAR ADAPTATIONS

Explain the difference between physiologic versus pathologic myocardial hypertrophy?

Explain why typical resistance training does not provide an adequate stimulus to improve cardiovascular status.

METABOLIC STRESS OF RESISTANCE TRAINING

Isometric and Weight-Lifting Exercise

Discuss the wisdom of a traditional weight training program to improve endurance capacity.

Circuit Resistance Training

Describe the circuit resistance method of training.

The energy cost of circuit resistance training equals what three other sport or recreational activities?

1.

2.

3.

BODY COMPOSITION ADAPTATIONS

Summarize the effects of DCER training on body composition.

MUSCLE SORENESS AND STIFFNESS

List four possible causes for delayed onset muscle soreness.

1.   3.

2.   4.

DOMS and Eccentric Muscle Action

Why do eccentric actions contribute to greater muscle damage and resulting soreness than concentric actions?

Cell Damage

List three specific cellular changes with DOMS

1.

2.

3.

Vitamin E Helps Reduce DOMS

Explain why vitamin E might alleviate delayed muscle soreness.

Spasm Theory

Discuss the spasm theory for DOMS.

Tear Theory

Outline the tear theory for DOMS.

Excess Metabolite Theory

Summarize the excess metabolite theory for DOMS.

Connective Tissue Damage Theory

What evidence indicates a contribution of connective tissue damage to DOMS?

Reducing DOMS

Describe one strategy to alleviate DOMS (and potential muscle damage.)

Practice Quiz

MULTIPLE CHOICE

1. Which is not used to measure a muscle's maximum force or tension:
   a. Tensiometry
   b. Dynamometry
   c. Plyometry
   d. One-repetition maximum
   e. a and c

2. To determine true "gender differences" in muscular strength, strength is most appropriately evaluated:
   a. Relative to muscle cross-sectional area
   b. On an absolute basis as total force exerted
   c. Relative strength per unit body mass
   d. Per unit percent body fat
   e. Relative to the lean-to-fat ratio

3. Which is not an appropriate explanation for muscle soreness and stiffness from resistance training?
   a. High accumulation of lactic acid
   b. Minute tears in muscle tissue
   c. Overstretching of muscle and connective tissue tears
   d. Repetitive eccentric muscle actions that accentuate fiber damage
   e. b and d

4. Type of training that utilizes inherent stretch-recoil characteristics of the neuromuscular system:
   a. Isometric
   b. Isokinetic
   c. Plyometric
   d. Periodization
   e. c and d

5. As movement velocity increases, greater torque results for individuals with this fiber-type composition:
   a. Predominantly fast-twitch
   b. Predominantly slow-twitch
   c. Equal percentage of fast and slow-twitch
   d. Predominantly type-I
   e. Predominantly SO

6. Which is false about a concentric muscle action?
   a. It occurs in dynamic activity
   b. The muscle lengthens while generating force
   c. Joint movement occurs as tension develops
   d. Raising a dumbbell from the extended to flexed elbow position
   e. b and de

7. Which is false about progressive resistance training?
   a. 3 to 9-RM represents most effective number of repetitions for strength increases
   b. Performing one set of an exercise improves strength than 2 or 3 sets
   c. Training 4 or 5 days a week produces less improvement than 2 or 3 days a week
   d. Fast rate of movement generates greater strength improvement than lifting at a slower rate
   e. None of the above

8. What minimum percent of a muscle's force-generating capacity should resistance be set to ensure strength increases?
   a. 10-20
   b. 20-40
   c. 60-80
   d. 80-90
   e. Greater than 90

9. Produces the greatest DOMS:
   a. Eccentric
   b. Concentric
   c. Isometric
   d. Isokinetic
   e. Plyometric

10. A factor that does not impact on muscle mass development and maintenance:
    a. Genetics
    b. Race
    c. Nutritional status
    d. Endocrine influence
    e. Pattern of nervous system activation

T R U E / F A L S E

1. _____ Men are usually stronger than women on an absolute basis.

2. _____ Increasing the load and decreasing the speed of muscular contraction creates overload.

3. _____ Heavy resistance training causes adaptations that enhance aerobic energy transfer.

4. _____ Increased muscle size with overload training occurs mainly by enlargement of individual muscle fibers.

5. _____ Isokinetic resistance training offers an effective alternative for combining the muscle-training benefits of resistance exercise with the cardiovascular benefits of continuous dynamic exercise.

6. _____ It is imprecise to term eccentric or concentric muscle actions as isotonic movements.

7. _____ Humans generate between 2 to 5 N of force per cm^2 of muscle cross section, regardless of gender.

8. _____ The lower back is not susceptible to significant injury with resistance training.

9. _____ Women develop less absolute increases in muscle girth with resistance training than men.

10. _____ Resistance training programs using moderate levels of concentric exercise can improve the strength of preadolescent children without adverse effects.

CROSSWORD PUZZLE

CHAPTER 15

ACROSS

5. increased number of muscle cells
7. force-_____ curve
9. _____ metric training is a system of static resistance training
10. first four letters of hormone that exerts ananabolic effect
11. weight training is considered a dynamic form of progressive resistance _____
12. CRT significantly increases the metabolic _____ of resistance training
14. organizing resistance training into phases of different types of exercise done at varying intensities and volumes for a specific time periods
17. 1-_____ is a measure of strength
20. another term for concentric-only movement
21. cable _____
22. differences in body size and composition mainly account for strength differences between _____

24. weight _____ is considered a dynamic form of PRE
25. first word in CRT
26. the weakest point in the joint range of motion is termed "sticking _____"
27. DOMS especially prevalent after this type of muscle action
28. first word in PRE

DOWN

1. enlargement of individual muscle fibers
2. muscle shortens while developing tension
3. a theory to explain muscle soreness
4. area of body susceptible to injury with improper resistance training
6. greatest gender difference when strength is expressed in this manner
8. rotational force on an isokinetic dynamometer
9. maximum force generated throughout movement at preset, constant speed

12. strength measurement device using steel spring
13. term for strength (resistance) training in which external resistance or weight does not change, but lifting (concentric) and lowering (eccentric) phases occur during each repetition (abbr.)
14. form of training that uses pre-stretch prior to force application
15. static resistance training popularized by German researchers in mid-1950s
16. initials describing muscle pain a day after unaccustomed exercise
18. this demand is relatively low with standard resistance training
19. first four letters of factor that accounts for rapid increase in strength early in training
23. Aabsolute strength/body mass is an expression of _____ strength
26. method for improving strength (abbr.)

Environment and Exercise

Define Key Terms and Concepts

1. Acute mountain sickness

2. ADH

3. Aldosterone

4. Alveolar P_{O_2} on Mt. Everest

5. Ambient P_{O_2} at 3048 m

6. Ambient P_{O_2} at sea level

7. AMS

8. Arterial desaturation

9. Break point for breath-hold

10. Cold acclimatization

11. Conduction

17. HACE

12. Convection

18. HAPE

13. Core temperature

19. Heat acclimatization

14. Diuretics

20. Heat cramps

15. Erythropoietin

21. Heat exhaustion

16. Evaporation

22. Heat illness

23. Heat stress index

24. Heat stroke

25. High altitude

26. Hyperthermia

27. Hyperventilation

28. Hypothalamus

29. Hypoxia

30. Insensible perspiration

31. Lactate paradox

32. Mean body temperature

33. Polycythemia

34. Radiation

35. Relative humidity (RH)

37. Wind chill index

36. WB-GT index

Study Questions

MECHANISMS OF THERMOREGULATION

THERMOREGULATION

How much does body temperature fluctuate during the day?

How much increase and decrease in core temperature can the body tolerate?

   Increase:

   Decrease:

THERMAL BALANCE

How many calories of energy can be generated each minute in sustained vigorous exercise?

Discuss the circulatory system's role in "fine-tuning" temperature regulation.

Body Temperature Measurement

List three common sites for body temperature measurement.

1.

2.

3.

Write the formula for determining mean body temperature from core and skin measurements.

HYPOTHALAMIC REGULATION OF BODY TEMPERATURE

Discuss the role of the hypothalamus in temperature regulation.

THERMOREGULATION IN COLD STRESS

Vascular Adjustments

What causes constriction of peripheral blood vessels during cold exposure?

Muscular Activity

In what way does shivering and physical activity contribute to thermoregulation during cold stress?

Hormonal Output

Name three hormones that increase the body's resting heat production.

1.

2.

3.

THERMOREGULATION IN HEAT STRESS

List the four ways the body loses heat.

1.   3.

2.   4.

Heat Loss by Radiation

Describe how radiation contributes to the body's heat loss.

Heat Loss by Conduction

Describe how conduction contributes to the body's heat loss.

Heat Loss by Convection

Describe how convection contributes to the body's heat loss.

Heat Loss by Evaporation

Describe how evaporation contributes to the body's heat loss.

Evaporative Heat Loss at High Ambient Temperatures

Which of the heat loss mechanisms become blunted when ambient temperature exceeds body temperature?

Heat Loss in High Humidity

List three factors that affect the total quantity of sweat vaporized from the skin.

1.

2.

3.

INTEGRATION OF HEAT-DISSIPATING MECHANISMS

Discuss how the circulatory system serves as the "workhorse" in temperature regulation.

Circulation

How much of the cardiac output passes through the skin to increase thermal conductance during extreme heat stress?

Evaporation

Discuss how circulation and evaporation work hand-in-hand in thermoregulation.

Hormonal Adjustments

What role does the pituitary gland play in the body's adjustment to heat stress?

EFFECTS OF CLOTHING ON THERMOREGULATION

Cold-Weather Clothing

Identify three characteristics of appropriate clothing for exercise in cold weather.

1.

2.

3.

Warm-Weather Clothing

Identify three characteristics of appropriate clothing for exercise in warm weather.

1.

2.

3.

Football Uniforms

Why do football uniforms pose a unique problem to thermoregulation during exercise in the heat?

NUTRITIONAL ASPECTS OF EXERCISE IN EXTREME ENVIRONMENTS

Give three dietary recommendations for recreational or expedition meal planning in environmental extremes.

1.

2.

3.

THERMOREGULATION AND ENVIRONMENTAL STRESS DURING EXERCISE

EXERCISE IN THE HEAT

Circulatory Adjustments

The body faces what two competitive demands when exercising in the heat?

1.

2.

Indicate the cardiac output differences in submaximum exercise in the heat compared with the cold.

Vascular Constriction and Dilation

Discuss whether prolonged reduction in blood flow to certain tissues can cause complications.

Maintaining Blood Pressure

In addition to redirecting blood to areas in great need, what other hemodynamic role does vasoconstriction serve during exercise in the heat?

CORE TEMPERATURE DURING EXERCISE

Plot the general relationship between esophageal temperature (Y-axis) and oxygen uptake as a percent of a person's $\dot{V}O_{2max}$ during graded exercise (X-axis). (Hint: Refer to Fig. 16.6 in your textbook.)

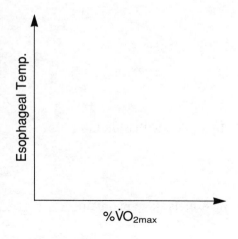

WATER LOSS IN THE HEAT

How much water loss occurs via sweating during moderate exercise at 90°F and 110°F?

90°F:

110°F:

Magnitude of Fluid Loss in Exercise

Discuss the following statement: "Excessive output of sweat in high humidity contributes little to the body's cooling."

Consequences of Dehydration

Indicate the effects of dehydration (use + for increase, − for decrease) on the following physiologic functions during exercise compared to conditions of normal hydration:

Variable effect of dehydration:

Rectal temperature:

Heart rate:

Sweat rate:

$\dot{V}O_{2max}$:

Fluid Loss in Winter Environments

List two factors responsible for water loss during cold-weather exercise.

1.

2.

Diuretic Use

List a unique negative effect of inducing dehydration with diuretics.

Water Replacement — Rehydration

The primary aim of fluid replacement during exercise is to maintain _____.

"The most effective defense against heat stress is _____."

Model for Rehydration

During heavy exercise at about 90°F and 50% relative humidity, give the recommended water intake per hour to avoid dehydration? (Hint: refer to Table 16.4 in your textbook.)

Electrolyte Replacement

Identify the sodium-conserving organ.

Why is it advisable to add a small amount of electrolytes to a rehydration fluid?

How much salt loss can occur during exercise in the heat?

FACTORS THAT IMPROVE HEAT TOLERANCE

Acclimatization

List five important physiologic adjustments during the process of heat acclimatization.

1.   4.

2.   5.

3.

Exercise Training

Give two important beneficial effects of exercise training on the body's ability to adjust to environmental heat stress.

1.

2.

Age

What is the general effect of age on thermoregulatory function?

Children

List three differences between children and adults during heat stress.

1.

2.

3.

Gender

What is the influence of gender on thermoregulatory function for the following variables?

Sweating:

Evaporative versus circulatory cooling:

Ratio of body surface area to volume:

Menstruation:

Sweating

List three differences between men and women in their sweating response.

1.

2.

3.

Evaporative Versus Circulatory Cooling

Explain how women show heat tolerance similar to men of equal aerobic fitness during identical exercise levels, despite a lower sweat output?

Body Surface Area-to-Mass Ratio

Why do women cool at a faster rate than males?

Menstruation

Discuss whether menstruation compromises a women's ability to sweat and dissipate heat during exercise.

Body Fat Level

Explain why fatal heat stroke occurs more frequently in the obese compared with the nonobese.

EVALUATING ENVIRONMENTAL HEAT STRESS

List four factors that determine the physiologic strain imposed by heat.

1.   3.

2.   4.

What is the most effective way to control heat-stress injuries?

Why is the WB-GT guide important?

Identify each of the specific components of the WB-GT index.

DBT:

WBT:

GT:

EXERCISE IN THE COLD

What is the general effect of exercising in cold water on energy metabolism during exercise?

What benefit does body fat provide during exposure to environmental cold stress?

Acclimatization to Cold

Give three examples of the body's limited adaptation to environmental cold stress.

1.

2.

3.

EVALUATING ENVIRONMENTAL COLD STRESS

List two warning signs of cold injury.

1.

2.

Wind Chill Index

List three environmental factors that contribute to increasing the Wind Chill Index.

1.

2.

3.

Respiratory Tract During Cold-Weather Exercise

Describe the effect of breathing cold air on the respiratory tract.

EXERCISE AT ALTITUDE

STRESS OF ALTITUDE

Indicate the partial pressure of oxygen (PO_2) in ambient air at each of the following terrestrial elevations (Hint: refer to Fig. 16.14 in your textbook.)

Sea level:

5000 feet:

10,000 feet:

15,000 feet:

20,000 feet:

Oxygen Loading at Altitude

Give two quantitative examples of the effect of various altitudes on $\dot{V}O_{2max}$ in comparison to values at sea level.

1.

2.

ACCLIMATIZATION

Present three broad guidelines for the length of time required to acclimatize to various altitudes. (Hint: refer to Table 16.6 in your textbook.)

1.

2.

3.

Immediate Adjustments

List two rapid physiologic responses that occur to compensate for the decrease in alveolar oxygen pressure at terrestrial elevations above 2300 meters.

1.

2.

Hyperventilation

Discuss why hyperventilation provides a desirable immediate first-line of defense against high altitude exposure.

FLUID LOSS

Why do people who exercise and train at altitude increase their risk for dehydration?

Accelerated Circulatory Response

Why does an increased cardiac output during submaximal exercise at altitude compared to sea level represent a beneficial physiologic adjustment?

Longer-Term Adjustments

List three important longer-term adjustments to prolonged stay at altitude.

1.

2.

3.

Describe the "lactate paradox."

Acid-Base Adjustment

Increased ventilation at altitude causes what change in alveolar P_{O_2}?

What is the result of carbon dioxide loss from body fluids during exposure to altitude?

Hematological Changes

Give two reasons for the increase in the blood's oxygen-carrying capacity during altitude acclimatization.

1.

2.

Describe the kidney's response to a prolonged stay at altitude.

Cellular Adaptations

List three long-term cellular adaptations to altitude exposure.

1.

2.

3.

ALTITUDE-RELATED MEDICAL PROBLEMS

Name the three medical conditions at high altitude that pose serious problems to ones overall health and safety.

1.

2.

3.

Acute Mountain Sickness

List three major symptoms of acute mountain sickness.

1.

2.

3.

High-Altitude Pulmonary Edema

How soon do HAPE symptoms appear following rapid ascent to high altitude?

List two significant symptoms of HAPE.

1.

2.

High-Altitude Cerebral Edema

List two significant symptoms of HACE.

1.

2.

EXERCISE CAPACITY AT ALTITUDE

Aerobic Capacity

Describe the relationship between $\dot{V}O_{2max}$ and increases in altitude.

Does the degree of physical conditioning prior to altitude exposure offer protection to the altitude-related decrease in $\dot{V}O_{2max}$?

Circulatory Factors

What causes the eventual reduction in exercise cardiac output in submaximal exercise during prolonged stay at altitude?

Give a possible cause for the reduced stroke volume observed at high altitude.

One- to three-mile running performance decreases by what magnitude during exposure to medium altitude?

ALTITUDE TRAINING AND SEA LEVEL PERFORMANCE

$\dot{V}O_{2max}$ on Return to Sea Level

List two effects of altitude training on subsequent performance on return to sea level.

1.

2.

Can Altitude Training Be Maintained?

Why should athletes periodically return from altitude to sea level for intensive training?

Altitude Training Versus Sea Level Training

Describe an experiment that evaluated the effect of aerobic training at altitude compared to equivalent training at sea level.

MULTIPLE CHOICE

1. Which is true about the hypothalamus?
   a. Receives input from receptors in the blood that are sensitive to changes in $\dot{V}O_2$ uptake
   b. "Thermostat" for temperature regulation
   c. Activated only during strenuous exercise
   d. Receives input from receptors in the blood that are sensitive to changes in acidity
   e. a and b

2. The "fine tuning" for temperature regulation:
   a. Hormonal adjustments
   b. Receptor adjustments
   c. Circulatory adjustments
   d. Ventilatory adjustments
   e. None of the above

3. This heat loss mechanism provides the major physiologic defense against overheating:
   a. Radiation
   b. Conduction
   c. Convection
   d. Evaporation
   e. Distillation

4. Which is true?
   a. One should wear one single bulky layer of clothing when exercising in the cold
   b. Thinner zones of trapped air next to the body corresponds to more effective insulation
   c. Wool or synthetics (polypropylene) insulate well and dry quickly
   d. The effectiveness of insulation increases when clothing becomes wet from sweating

5. Which is false about circulatory adjustments in submaximal exercise when exercising in cold weather:
   a. Stroke volume of the heart is higher in the heat
   b. Cardiac output is the same whether exercising in the cold or heat
   c. Heart rate is higher in the heat
   d. Systolic blood pressure is the same whether exercising in the cold or heat
   e. a and d

6. A consequence of pre-exercise dehydration:
   a. Increase in rectal temperature
   b. Decrease in sweat rate
   c. Decrease in $\dot{V}O_{2max}$
   d. a and b
   e. All of the above

7. Which has the least effect on heat tolerance?
   a. Acclimatization
   b. Training
   c. Age
   d. Fatness
   e. Heredity

8. Which is true?
   a. Humans possess less capacity for adaptation to cold than heat exposure
   b. Humans possess more capacity for adaptation to cold than heat exposure
   c. Humans possess similar capacities for adaptation to cold and heat exposure
   d. Humans possess less capacity for cold than heat adaptation only at sea level ambient pressure
   e. Humans possess more capacity for cold than heat adaptation only at sea level ambient pressure

9. The immediate physiologic adjustments to altitude begins with:
   a. Decrease in alveolar P_{CO_2}
   b. Increase in alveolar P_{CO_2}
   c. Decrease in alveolar P_{O_2}
   d. Decrease in alveolar P_{O_2}
   e. None of the above

10. Which is not a long-term adjustment to an increase in altitude?
    a. Increase in submaximal cardiac output
    b. Acid-base readjustment
    c. Increase in blood's oxygen-carrying capacity
    d. Increase in capillary concentration in skeletal muscle
    e. b and d

T R U E / F A L S E

1. _____ The surface area exposed to the environment represents the most important factor for evaporative heat loss.

2. _____ Fifteen to 25% of cardiac output passes through the skin in extreme heat stress during rest.

3. _____ For some athletes, fluid loss from sweating represents 6 to 10% of body mass.

4. _____ Localized "cold treatments" represent the most effective defense against heat stress.

5. _____ Exercise training causes sweating to begin at a lower body temperature.

6. _____ Women possess less heat-activated sweat glands per unit skin area than men.

7. _____ Fatal heat stroke occurs 3.5 times more frequently in young adults who are overweight than normal-weight individuals.

8. _____ Inhalation of cold ambient air can significantly damage the respiratory passages.

9. _____ Reduced loading of hemoglobin with oxygen occurs at high altitudes.

10. _____ Aerobic capacity decreases linearly with increases in altitude.

CROSSWORD PUZZLE

ACROSS

1. pulmonary disorder at altitude (abbr.)
4. percentage of _____ in air on Mt. Everest is 20.9
5. heat stress index that uses temperature, humidity, and radiant heat (abbr.)
8. thirty to 40% of heat is lost from this body area
13. this hormone stimulates production of red blood cells
14. this category of activities is negatively affected at high altitude
16. group at high risk when working in the heat
18. a dietary deficiency in this mineral may hinder acclimatization (abbr.)
20. erug or substance that promotes body water loss
25. regulation of body temperature
31. first 3 letters of gland that releases ADH
32. maladaptation to high altitude that affects central nervous system (abbr.)
34. deeper body tissues
37. hormone released from adrenal cortex to increase sodium reabsorption
39. acts on kidneys to conserve water during heat stress (abbr.)
40. women divers of Korea and southern Japan
41. moisture content of air
42. this tissue "wins out" in the competition for blood in hot weather exercise
44. adaptive response that improves tolerance to altitude hypoxia
45. _____ balance is achieved when heat loss = heat again

DOWN

2. the length of the altitude acclimatization depends on this factor
3. this chamber simulates altitude
6. a calorigenic hormone
7. this group demonstrates a lower rate of sweating and a higher core temperature during heat stress
9. highest mountain in the world
10. this quality of ambient gaseous environment decreases progressively with increasing altitude
11. index of ambient cold
12. _____ perspiration
15. electrolyte, when added in small amounts to fluid may sustain thirst drive (abbr.)

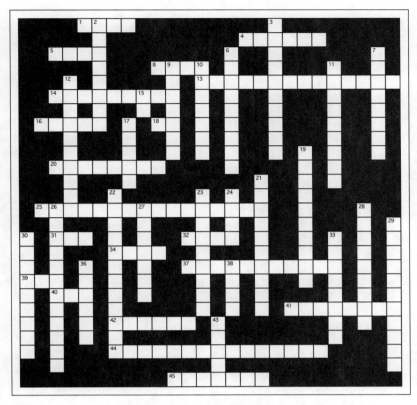

17. upper limit of beneficial percentage increase in hemoglobin concentration at altitude is about _____ percent
19. heat transfer involving air or fluid movement
21. direct molecular transfer of heat through a liquid, solid, or gas
22. abnormally high concentration of red blood cells
23. cold weather clothing that covers nose and mouth
24. elite marathoners frequently experience fluid losses in excess of _____ liters during competition
26. coordinating center for temperature regulation
27. sweat glands are of this type
28. concentration of this iron-containing protein compound in muscle increases with long-term altitude acclimatization
29. this macronutrient is well tolerated and potentially beneficial at high altitude
30. a condition of abnormally high pH
33. another term for perspiring
35. _____ fluid volume decreases during the first several days of altitude exposure
36. the major portion of sweat is drawn from this body fluid
38. poor indicator of core temperature after strenuous exercise
43. cools body 2 to 4 times faster than air at same temperature

Ergogenic Aids

Define Key Terms and Concepts

1. Amphetamine

2. Anabolic steroid

3. Anti-cortisol producing compounds

4. Anxiolytic effect

5. Autologous transfusion

6. Benzedrine

7. Bicarbonate loading

8. Blood doping

9. Caffeine

10. Carnitine

11. Chromium

12. Clenbuterol

13. CoQ$_{10}$

14. Cortisol

15. Creatine

16. Creatine loading

17. C$_r$H$_2$O

18. DHEA

19. Ergogenic aid

20. Ergolytic effect

21. Erythropoietin

22. General warm-up

23. Glutamine

24. Gynecomastia

25. Homologous transfusion

26. Hormonal blood boosting

27. Hyperoxic gas mixtures

28. L-Carnitine

29. MAST

30. MORA

31. Pangamic acid

32. PCr

33. Phosphate Loading

34. Phosphatidylserine

35. Picolinate

38. Stacking

36. Pyramiding

39. Vitamin B$_{15}$

37. Sodium Citrate

Study Questions

PHARMACOLOGIC AND NUTRITIONAL AGENTS

List three possible mechanisms that explain how nutritional and pharmacologic agents might enhance performance.

1.

2.

3.

List four substances currently banned by the IOC.

1.   3.

2.   4.

ANABOLIC STEROIDS

Structure and Action

Define anabolic steroids and give their mode of action.

Definition:

Mode of action:

Estimates of Steroid Use

Approximately how many high school students have used steriods?

Effectiveness of Anabolic Steroids

List two sources of confusion concerning the effectiveness of anabolic steroids as an ergogenic aid?

1.

2.

Dosage an Important Factor

Would weekly injections of 600 mg testosterone be considered a low or high dose?

Risks of Steroid Use

List five potentially harmful side effects of orally administered steroids.

1.   4.

2.   5.

3.

ACSM Position Statement on Anabolic Steroids

List three major points of the American College of Sports Medicine's position stand on use of anabolic steroids.

1.

2.

3.

Steroid Use and Life-Threatening Disease

Describe how steroid use can be life-threatening.

Steroid Use and Plasma Lipoproteins

Indicate the effects of anabolic steroids on blood lipids?

Steroid Use by Females

List two possible negative effects of anabolic steroids unique to females.

1.

2.

ANDROSTENEDIONE: A LEGAL SUPPLEMENT IN SOME SPORTS

List one professional sport that does not ban the use of androstenedione.

Action and Effectiveness

Describe androstenedione.

CLENBUTEROL: ANABOLIC STEROID SUBSTITUTE

Describe the current medical use of Clenbuterol and its proposed ergogenic effects.

Medical use:

Ergogenic effect:

Potential Negative Effect on Muscle Function

Describe two potential negative effects of Clenbuterol use.

1.

2.

GROWTH HORMONE: THE NEXT MAGIC PILL?

What is the normal medical use for human growth hormone and what risks exist from its unsupervised use?

Medical use:

Risks:

DHEA: NEW DRUG ON THE CIRCUIT

Where in the body is DHEA synthesized?

DHEA Safety

Describe DHEAs use and possible negative side effects.

Use:

Side Effects:

AMPHETAMINES

List three effects of amphetamines on normal physiologic function.

1.

2.

3.

Dangers of Amphetamines

List four potentially dangerous side effects of amphetamine use in athletics.

1.   3.

2.   4.

Amphetamine Use and Athletic Performance

Summarize the general research findings concerning the ergogenic effects of amphetamines.

CAFFEINE

Ergogenic Effects

Explain the main facilitating effect of caffeine as an ergogenic aid.

Proposed Mechanism for Ergogenic Action

How does caffeine exert its ergogenic effect on endurance performance?

Endurance Effects Often Inconsistent

List two factors that may blunt the ergogenic benefit of caffeine for endurance exercise.

1.

2.

Effects on Muscle

What is the proposed ergogenic effect of caffeine on neuromuscular function?

Warning

List three negative side effects of caffeine.

1.

2.

3.

ALCOHOL

Is alcohol classified as a fat, protein, or carbohydrate?

Give the approximate number of individuals who abuse alcohol in the U.S.

Use Among Athletes

What is the best predictor of an athlete's alcohol use?

Psychological and Physiological Effects

Describe alcohol's ultimate effect on the body.

List two potential positive effects of alcohol.

1.

2.

Alcohol and Fluid Replacement

Give two reasons why alcohol accelerates dyhydration during exercise in the heat.

1.

2.

How does alcohol impede rehydration?

PANGAMIC ACID

Give another name for pangamic acid.

Indicate the proposed ergogenic benefit of pangamic acid and what are the research findings concerning this benefit?

Proposed benefit:

Research findings:

BUFFERING SOLUTIONS

How does ingestion of a pre-exercise buffering solution enhance short-term exhaustive exercise performance?

Effects Relate to Dosage and Degree of Exercise Anaerobiosis

What dose of bicarbonate facilitates H^+ efflux from the cell?

PHOSPHATE LOADING

List three effects of pre-exercise phosphate supplementation.

1.

2.

3.

Give one reason for inconsistencies in research findings concerning phosphate loading.

ANTI-CORTISOL PRODUCING COMPOUNDS

Glutamine

List two possible beneficial effects of glutamine.

1.

2.

Phosphatidylserine

Present the potential positive effects of consuming anti-cortisol compounds?

Other than its use as an anti-cortisol compound, what positive effects are reported for glutamine?

Normal Cortisol Release: A Bad Response?

Discuss whether cortisol provides any beneficial effects.

CHROMIUM

Describe chromium's normal function in the body.

Alleged Benefits

List three alleged benefits of chromium supplementation.

1.

2.

3.

Not Without a Potential Downside

List two potential negative side-effects of chromium supplementation.

1.

2.

CREATINE

List two rich sources of creatine in the diet.

1.

2.

Important Component of High-Energy Phosphates

Complete the equation:

$$PCr + ADP \longrightarrow$$

Give three important functions of PCr in the body.

1.

2.

3.

Documented Benefits

List two possible performance-enhancing benefits of creatine supplementation.

1.

2.

Effects on Body Mass

List two possible reasons why creatine exerts an ergogenic effect.

1.

2.

Creatine Loading

Describe the typical creatine loading procedure.

Caffeine Blunts Creatine's Effect

How does caffeine negate creatine's effect?

PHYSIOLOGIC AGENTS
RED BLOOD CELL REINFUSION

Give another name for red blood cell reinfusion.

How It Works

Outline the general procedure of red blood cell reinfusion for ergogenic purposes.

Does It Work?

Give a quantitative example of how the addition of red blood cells increases the blood's oxygen-carrying capacity.

Describe a potential danger of blood doping.

What are the physiologic and performance benefits of red blood cell reinfusion?

A New Twist—Hormonal Blood Boosting

Indicate the physiologic function of erythropoietin and discuss its danger as an ergogenic aid.

Physiologic function:

Potential danger:

WARM-UP (PRELIMINARY EXERCISE)

Psychologic Considerations

Give the psychologic basis for warming-up.

Physiologic Considerations

List four physiologic mechanisms by which warm-up could potentially improve subsequent exercise performance.

1.   3.

2.   4.

Effects on Performance

What are the ergogenic benefits of warming-up and give a recommendation concerning its use.

Ergogenic benefits:

Recommendation:

Warm-up and Sudden Strenuous Exercise

What research evidence justifies the potential cardiovascular benefits of warm-up prior to sudden, strenuous exercise?

BREATHING HYPEROXIC GAS

Give the physiologic rationale for breathing oxygen-enriched gas mixtures prior to, during, or in recovery from strenuous exercise.

Pre-Exercise

Explain why pre-exercise breathing of a hyperoxic gas mixture provides no ergogenic effect.

During Exercise

Outline the research findings for the ergogenic benefits of breathing hyperoxic gas during submaximal and maximal aerobic exercise.

Submaximal exercise:

Maximal exercise:

Why does breathing hyperoxic gas during exercise provide an ergogenic effect in aerobic exercise?

In Recovery

What effect does breathing hyperoxic gas have on recovery from strenuous exercise?

Practice Quiz

MULTIPLE CHOICE

1. Hyperoxic gas breathed during exercise:
   a. Enhances endurance activities at high altitude
   b. Increases running economy at high altitude
   c. Enhances anaerobic activities at high altitude
   d. Facilitates high-altitude acclimatization
   e. a and d

2. Which is false about anabolic steroids?
   a. They function similar to the male hormone estrogen

   b. They bind with specific receptor sites on muscle
   c. They contribute to male secondary sex characteristics
   d. They contribute to sex differences in muscle mass and strength
   e. None of the above

3. Which is not a consequence of taking anabolic steroids?
   a. Impaired normal testosterone endocrine function
   b. Increased concentration of estradiol

   c. Gynecomastia

   d. Decreased secretion of stress-related hormones

   e. Acne

4. Which is true about steroid use and plasma lipoproteins?
   a. Lowers high-density lipoprotein cholesterol
   b. Lowers low-density lipoprotein cholesterol
   c. Lowers total cholesterol
   d. b and c
   e. None of the above

5. Which is not a function of human growth hormone?
   a. Stimulates amino acid uptake and protein synthesis by muscle
   b. Increases fat breakdown
   c. Prevents lactic acid buildup in muscle
   d. Decreases carbohydrate utilization
   e. b and c

6. Glutamine is a:
   a. Nonessential amino acid
   b. Polyunsaturated fatty acid
   c. Form of amphetamine
   d. Component of buffering solutions
   e. a and d

7. Which is true concerning pangamic acid?
   a. Commonly known as vitamin B_{10}
   b. Increases cell's efficiency to use protein for fuel
   c. Reduces lactic acid buildup and enhances endurance performance
   d. Serves no particular nutritive role for the body
   e. a and d

8. Physiologic adjustments in response to red blood cell reinfusion (blood doping):
   a. Red blood cell count and hemoglobin levels increase
   b. Added blood volume contributes to larger maximal cardiac output
   c. Blood's oxygen-carrying capacity increases
   d. a and c
   e. All of the above

9. Benefits attributed to "warming-up":
   a. Increased speed of muscle contraction and relaxation
   b. Facilitated oxygen release to muscles
   c. Facilitated nerve transmission
   d. Increased blood flow through active tissues
   e. All of the above

10. Purported benefits of chromium supplementation:
   a. Fat burner
   b. Muscle builder
   c. Increases lean body mass
   d. b and c
   e. All of the above

TRUE / FALSE

1. _____ "Stacking" refers to anabolic steroid use every other week.

2. _____ Anabolic steroids increase maximal aerobic power.

3. _____ Human growth hormone can cause gigantism in children and acromegalia in adults.

4. _____ The ergogenic effect of caffeine occurs through facilitated use of fat as fuel for exercise.

5. _____ Bicarbonate loading does not adversely affect the body.

6. _____ Possible benefits of increased intramuscular PCr include faster ATP turnover rate to maintain power output during short-term, all-out effort.

7. _____ Caffeine blunts the ergogenic effect of creatine supplementation.

8. _____ Clenbuterol substitutes for anabolic steroids.

9. _____ Breathing high concentrations of oxygen after strenuous exercise greatly increases oxygen transport by hemoglobin.

10. _____ The side effects of amphetamines include drug dependency, headache, and confusion.

CROSSWORD PUZZLE

ACROSS

1. tissue building
3. infusing a person's own blood
5. type of warm up involving calisthenics and stretching
10. progressively increasing steroid dose in oral and injectable form
12. first 4 letters of chemicals known as "downers"
13. dehydroepiandroster one (abbr.); relatively weak steroid hormone synthesized primarily by primates in the adrenal cortex from cholesterol
15. progressively increasing the drug dosage during six- to 12-week cycles
17. ergogenic technique that enhances anaerobic capacity
20. erythrocyte (abbr.)
22. _____ trophic hormone is another term for human growth hormone
24. hydrocortisone, decreases amino acid transport into cells; blunts anabolism and stimulates protein breakdown to its building-block amino acids in all cells except the liver; also serves as an insulin antagonist by inhibiting glucose uptake and oxidation
25. _____ ecomastia is palpable breast tissue in male steroid abusers
28. something that enhances exercise capacity
29. mimics action of catecholamines
31. meat, poultry, and fish provide rich sources of this substance; the body synthesizes only about 1 g of this nitrogen-containing organic compound daily, primarily in kidney, liver, and pancreas from the nonessential amino acids arginine, glycine, and methionine
32. shown to increase skeletal muscle mass and slow fat gain in certain experimental animals to counter the effects of aging, immobilization, malnutrition, and tissue-wasting pathology; steroid substitute

DOWN

1. "pep pill"
2. methylxanthine
4. facilitates the influx of long-chain fatty acids into the mitochondrial matrix where they enter Beta-oxidation during energy metabolism
6. hormone normally produced by the kidneys;

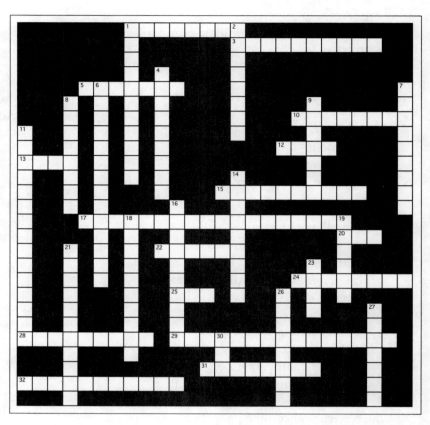

stimulates bone marrow to produce red blood cells; combats anemia in patients with severe kidney disease
7. dangerous side effects that actually impair performance
8. _____ acid is another name for "Vitamin B-15"
9. anabolic _____ is a drug that functions similar to testosterone
11. an intermediate or precursor hormone between DHEA and testosterone, aids the liver to synthesize other biologically active steroid hormones
14. oxygen enriched gas
16. transfusion that infuses a type-matched donor's blood
18. reduces tension and anxiety
19. _____ hormone may replace anabolic steroids as an abused ergogenic aid
21. red blood cell reinfusion
23. _____ cythemia is increased concentration of red blood cells
26. trace mineral that serves as a cofactor for potentiating insulin function; touted as a "fat burner" and "muscle builder"
27. last 7 letters of kidney hormone that stimulates bone marrow to produce RBCs
30. creatine combined with phosphate (abbr.)

Body Composition: Components, Assessment, and Human Variability

1. 495/density − 450 =

2. Amenorrhea

3. Archimedes' principle

4. Bioelectric impedance analysis

5. Bod Pod

6. Body mass index

7. Body volume

8. Buoyancy

9. Computed tomography

10. Density

11. Desirable body mass

17. Skinfold caliper

12. DEXA

18. Fat-free body mass

13. Dr. Albert Behnke

19. Gender-specific essential fat

14. Dual-energy x-ray absorptiometry

20. Generalized equations

15. Essential fat

21. Girth measurements

16. Fat mass

22. Height-weight tables

23. Hydrostatic weighing

24. Lean body mass

25. Minimal weight

26. Minimal wrestling mass

27. MRI

28. NIR

29. Oligomenorrhea

30. Overweight

31. Reference man

32. Reference woman

33. Residual lung volume

34. Sex-specific fat

35. Specific gravity

39. Underweight

36. Storage fat

40. Visceral fat

37. Subcutaneous skinfolds

41. William Siri

38. Ultrasound

Study Questions

Describe the fundamental discovery concerning body composition by Dr. A. Behnke made in the early 1940s?

GROSS COMPOSITION OF THE HUMAN BODY

List three major structural components of the human body and their percentage as represented by the reference man and woman. (Hint: Refer to Figure 18.2 in your textbook.)

Structural Component	Reference Man	Reference Woman
1.		
2.		
3.		

Reference Man and Reference Woman

Compare body mass, stature, total fat, and storage and essential fat for the reference man and woman.

	Reference Man	Reference Woman
Stature, cm:		
Body mass, kg:		
Total fat, kg:		
Total fat, %:		
Storage fat, kg:		
Storage fat, %:		
Essential fat, kg:		
Essential fat, %:		

Essential and Storage Fat

Essential Fat

Give the function and location of essential fat and sex-specific essential fat in humans.

	Function	Location
Essential fat:		
Sex-specific fat:		

Storage fat

Give the function and location of storage fat in humans.

Function:

Location:

Fat-Free Body Mass and Lean Body Mass

Define and indicate the components of lean body mass in men.

Definition:

Components:

What is the suggested "healthy" lower level of percent body fat in males?

Minimal Body Mass

Define and indicate the components of the minimal weight in females.

Definition:

Components:

What is the suggested "healthy" lower level of percent body fat in females?

Underweight and Thin

What precisely is meant by the terms "underweight" and "thin?"

Underweight:

Thin:

LEANNESS, EXERCISE, AND MENSTRUAL IRREGULARITY

What is the association between menstrual irregularity and body fat?

Describe the lower limit of body fat required to maintain normal menstrual function?

Leanness Not the Only Factor

List four factors associated with menstrual dysfunction.

1.   2.

3.   4.

Delayed Onset of Menstruation and Cancer Risk

Discuss the relationship between delayed onset of menstruation and cancer risk and indicate the proposed cause for this association?

Relationship:

Proposed Cause:

METHODS TO ASSESS BODY SIZE AND COMPOSITION

List two general procedures to evaluate body composition.

1.

2.

Describe a major limitation of the direct method of body composition assessment in humans.

List three indirect procedures commonly used to assess body composition.

1.

2.

3.

State Archimedes principle of water displacement.

How is Archimedes principle of water displacement used to evaluate body composition?

Determining Body Density

Complete the formula: Specific gravity = _____ ÷ _____.

Calculate the approximate body volume of a person weighing 50 kg who weighs 2 kg when submerged underwater.

Computing Percent Body Fat, Fat Mass, and Fat-Free Body Mass

Write the Siri equation to compute percent body fat from body density.

Compute the percent body fat of a person whose body density equals 1.0742 g/cc.

Give the equation and compute the fat mass for a person weighing 63.4 kg with body fat of 10.8%.

   Equation:

   Fat mass:

Give the equation and compute the fat-free body mass for a person weighing 63.4 kg with 6.85 kg of fat.

   Equation:

   Fat-free body mass:

Limitations and Errors in Hydrostatic Weighing

List two possible sources of error when using the generalized density value of $1.10 \text{ g} \cdot \text{cc}^{-3}$ for fat-free tissue.

1.

2.

Body Volume Measurement

Write the equation to compute body volume by hydrostatic weighing.

EXAMPLES OF CALCULATIONS

Compute total percent body fat for a black male with a body mass of 110 kg, underwater weight of 3.5 kg, RLV of 1.2 L, and water correction factor of 0.996. (Remember to use the correct percent body fat equation.)

A WORD ABOUT RESIDUAL LUNG VOLUME

Why must residual lung volume be accounted for when computing body volume by hydrostatic weighing?

Body Volume Measurement by Air Displacement

Explain the principle underlying the use of the BOD POD for body volume determinations.

Skinfold Measurements

The close relationship between what three variables provide the rationale for using skinfold measurements to predict body composition.

1.

2.

3.

The Caliper

After applying the caliper to the skin, how soon should readings be made?

Sites

List five common sites for measuring skinfolds and their anatomic location.

<u>Skinfold site</u>   <u>Anatomic location</u>

1.

2.

3.

4.

5.

Use of Skinfold Data

Give two ways to use skinfold scores when assessing body composition.

1.

2.

Girth Measurements

USEFULNESS OF GIRTH MEASUREMENTS

List two advantages of girth measurements over skinfolds to assess body fat.

1.

2.

PREDICTING BODY FAT FROM GIRTHS

Compute percent body fat and fat-free body mass for a 35-yr-old woman who weighs 70.4 kg with the following girth measurements: abdomen = 82.6 cm; right thigh = 57.8 cm; right calf = 44.5 cm.

   Percent body fat:

Fat-free body mass:

Bioelectrical Impedance Analysis (BIA)

How accurate is the BIA method compared with the skinfold method to assess body fat?

Consider Level of Hydration and Ambient Temperature

Why does hydration level affect BIA measurements?

Dual-Energy X-ray Absorptiometry

Describe the relationship between percent fat determined by DEXA and percent fat determined by densitometry.

BODY MASS INDEX

Write the formula to compute body mass index.

Why is the BMI an important component in a medical evaluation?

Limitations of BMI for Athletes

Give one limitation of the BMI and its importance to athletes?

Limitation:

Importance to athletes:

AVERAGE VALUES FOR BODY COMPOSITION

Give "average" values for percent body fat for young and older-age adult men and women. (Hint: Refer to Table 18.5 in your textbook.)

Young men: Older men:

Young women: Older women:

Representative Samples Lacking

Give an explanation for the observed change in percent body fat with increasing age.

Determining Goal Body Weight

Write the equation to compute desirable body mass.

A 20-year-old man weighs 89 kg with 22% body fat. If this man reduces body fat to a desired 12% level, determine (a) his new body mass, and (b) total fat mass lost? Assume all weight loss represents fat.

New body mass:

Total fat loss:

BODY COMPOSITION OF CHAMPION ATHLETES

Long-Distance Runners

Provide representative values for body mass, stature, and percent body fat for elite male and female distance runners and untrained counterparts.

	Body Mass	Stature	Percent Body Fat
Male Runners:			
Untrained Males:			
Female Runners:			
Untrained Females:			

High School Wrestlers

Give the recommendation of the American College of Sports Medicine concerning the lowest acceptable level of body fat for safe wrestling competition.

Male and Female Weight Lifters and Body Builders

Provide percent body fat and excess muscle of male competitive body builders, weight lifters, and power weight lifters.

	Percent Body Fat	Excess Muscle
Competitive body builders:		
Weight lifters:		
Power weight lifters:		

Provide physical characteristics for competitive female body builders and the reference woman:

	Body Mass	Stature	Percent Body Fat	Fat-free Body Mass
Body Builder:				
Reference Woman:				

Provide average percent body fat levels for these groups of male athletes:

Body builders:

Olympic weight lifters:

Power weight lifters:

What is the FFM/FM ratio for competitive female body builders? The reference female?

Practice Quiz

MULTIPLE CHOICE

1. It is possible to compute all of the following from underwater weighing except:
   a. Body density
   b. Percent body fat
   c. Fat-free body mass
   d. Approximate number of fat cells
   e. Body volume

2. Main limitation of height-weight tables:
   a. Do not account for differences in bone density
   b. Do not consider age
   c. Give no indication of body composition
   d. Based on old normative data
   e. c and d

3. According to Behnke's model for the reference man and reference woman:
   a. A woman jeopardizes health if body fat falls below 22%
   b. A man jeopardizes health if body fat falls below 8%
   c. The average quantity of essential fat for men equals about 3% of body mass
   d. The average quantity of storage fat for women equals about 26% of body mass.
   e. All of the above

4. Lean female athletes often experience secondary amenorrhea; this indirectly verifies Behnke's concept of:
   a. Minimal weight
   b. Storage fat
   c. Energy drain
   d. Lean body mass
   e. a and c

5. The average percent body fat for young adult men and women, respectively, equals:
   a. 12%, 22%
   b. 15%, 25%
   c. 18%, 28%
   d. 20%, 30%

6. Percentage of body fat cited as the critical level for the onset of menstruation:
   a. 15%
   b. 17%
   c. 19%
   d. 21%
   e. 25%

7. Subtract _____ from an individual's actual body volume when calculating body volume.
   a. Residual lung volume
   b. Vital capacity
   c. Expiratory reserve volume
   d. Inspiratory reserve volume
   e. Functional residual capacity

8. A drawback in using skinfolds to predict percent body fat:
   a. Experience required in taking the measurements
   b. Skinfold thickness for obese people often exceeds the caliper's jaw opening
   c. Particular caliper used may contribute to measurement error
   d. Prediction equations contribute to prediction error
   e. All of the above

9. Factors that affect accuracy of BIA except:
   a. Hyperhydration
   b. Skin temperature
   c. Dehydration
   d. Person's height
   e. Fat cellularity

10. Lower acceptable limit of percent body fat for wrestlers?
    a. 3 to 5
    b. 5 to 7
    c. 7 to 9
    d. 9 to 11
    e. 12 to 15

TRUE / FALSE

1. _____ Body mass, lean body mass, and percentage fat represent the three major structural components of the body.

2. _____ The error in predicting percent body fat from skinfold or girth equations generally lies within 2.5 to 4.0% fat units from percent body fat determined by hydrostatic weighing.

3. _____ Desirable body mass computes as fat-free body mass minus percent fat desired.

4. _____ The existing equations to calculate body composition from body density in whites tend to overestimate the fat-free body mass for blacks.

5. _____ The essential fat percentage equals 8% for men and 15% for women.

6. _____ Girth measurements provide meaningful information about body fat and its distribution.

7. _____ As one ages, a greater quantity of fat deposits as subcutaneous fat than internal fat.

8. _____ Magnetic resonance imaging provides valuable information about the body's tissue compartments.

9. _____ Changes in body composition with age partly reflect decreases in bone density.

10. _____ The most striking compositional characteristic of the female body builder is her large lean-to-fat ratio of 4:1.

CROSSWORD PUZZLE

ACROSS

6. these equations consider age when predicting body fat
7. mass per unit volume
9. instrument to measure skinfolds
10. navy physician who pioneered work on body composition
11. absence of menses
12. body mass devoid of extractable fat (abbr.)
14. dimension of body build characterized by a rounded body shape and predominance of fat
16. lung volume that is subtracted to determine body volume
22. internal fat in relation to subcutaneous fat increases with this variable
23. ratio of an object's mass to the mass of equivalent volume of water
25. risk for this disease is lower in female athletes with delayed onset menstruation
27. reference _____ averages 15% body fat
28. body composition technique that measures electrical flow in the body (abbr.)
29. muscle mass plus a small percentage of essential fat stored chiefly within the CNS, bone marrow, and internal organs (abbr.)
32. this component of the body's lipids serves as an energy depot
33. this strength measure is not a good indication of excessive lean body mass loss in wrestlers
34. 0.90 g/cc is the density of this tissue
35. percentage fat proposed as the lowest acceptable level for safe wrestling competition
36. body's external surface area (abbr.)
38. adding this to a weight loss program favorably affects the composition of the weight lost
39. lower limit of body weight for women that includes about 14% essential fat
40. _____ weighing is a technique to assess body volume
41. dimension of body build characterized by a tall, thin, and linear body

DOWN

1. Condition where body mass is greater than some standard, usually the mean body mass for a given stature
2. body composition assessment technique using high frequency sound
3. dimension of body build characterized by well developed musculature

4. athletes that perform three different types of exercise in succession
5. too little fat for frame-size
8. reference _____ averages 25 to 27% body fat
13. condition where body mass is less than some standard, usually the average body mass for a given stature
15. irregular menstrual cycle
17. body typing that assigns a characteristic shape and physical appearance
18. there are _____ differences in the density of the body's fat-free component
19. elite athletes in this sport have relatively short arms and legs for their stature
20. lipids in bone marrow heart, lungs, liver, spleen, kidneys, intestines, muscles, and lipid rich tissues in the CNS
21. body composition technique using magnetic imaging (abbr.)
22. discovered principle of water displacement
24. the physique of male and female triathletes is most similar to these single sport athletes
26. body composition technique using infrared light (abbr.)
30. body composition technique using roentgenograms
31. dual x-ray absorptiometry (abbr.)
36. body mass, kg ÷ stature, m2 (abbr.)
37. another term for circumference measurements

Obesity, Exercise, and Weight Control

1. 3500 kcal

2. Adipocyte

3. Android obesity

4. Creeping obesity

5. Criterion for overfatness

6. Desirable waist-to-hip girth ratio

7. Dose-response

8. Ectomorphy

9. Endomorphy

10. Energy balance equation

11. Familial aggregation

12. Fat cell biopsy

13. Fat cell hyperplasia

14. Fat cell hypertrophy

15. Gynoid obesity

16. Health risks of obesity

17. High-protein diets

18. Ketogenic diets

19. Lean-to-fat ratio

20. Leptin

21. Lipoprotein lipase

22. Making weight

23. Mesomorphy

24. Ob gene

25. Obesity

26. Obesity-producing environment

27. Patterning of adipose tissue

28. Pedometer

29. Ponderal (mass) equivalents

30. Preadipocyte pool

31. Protein-sparing modified fast

32. Recovery "afterglow"

33. Regional fat distribution

34. Setpoint theory

35. Somatotyping

36. Spot reduction

37. Standards for overfatness

38. Visceral adipose tissue

39. VLCD

40. Waist-to-hip girth ratio

41. Yo-yo effect

Study Questions

OBESITY
OBESITY: A LONG-TERM PROCESS

Discuss the trend for weight gain in adult men and women as they age.

NOT NECESSARILY OVEREATING

List five factors that predispose a person to excessive weight gain.

1.   4.

2.   5.

3.

Genetics Play a Role

How much of the variation in weight gain among individuals can be accounted for by genetic factors?

A Mutant Gene?

Describe the role of the mutant "obese" gene in the obesity development.

Physical Activity: An Important Component

Describe the relationship between body fat and age.

HEALTH RISKS OF OBESITY

Does excess body weight or excess body fat relate more strongly to heart disease risk?

HOW FAT IS TOO FAT?

List three criteria for evaluating a person's level of body fatness.

1.

2.

3.

Percent Body Fat

What percent body fat level indicates borderline obesity in adult men and women?

  Men:

  Women:

Fat Patterning

Identify the two patterns of excessive fat distribution among obese men and women.

1.

2.

Give an objective standard for establishing male and female-pattern obesity.

  Male:

  Female:

Fat Cell Size and Number

Give two cellular changes that would increase the body's quantity of adipose tissue.

1.

2.

Fat Cell Size and Number in Normal and Obese Adults

List average values for fat cell size and fat cell number in obese and nonobese adults.

	Obese	Nonobese
Cell size:		
Cell number:		

Fat Cell Size and Number After Weight Loss

What happens to fat cell size and fat cell number when adults lose weight?

  Cell size:

  Cell number:

Discuss the evidence concerning the possibility of "curing" fat cell hyperplasia.

Fat Cell Size and Number After Weight Gain

Describe changes to fat cell size and number with moderate weight gain in adults?

    Fat cell size:

    Fat cell number:

New Fat Cells Can Develop

Describe changes in fat cell number during massive weight gain in adults?

ACHIEVING OPTIMAL BODY COMPOSITION THROUGH DIET AND EXERCISE

Briefly discuss the prognosis for successful weight loss.

THE ENERGY BALANCE EQUATION

Write the energy balance equation.

Personal Assessment

Energy Intake

List two reasons for keeping food intake records during a weight loss program.

1.

2.

Energy Output

What is the main factor that contributes to increased daily energy output?

DIETING TO TIP THE ENERGY BALANCE EQUATION

Describe a prudent approach to unbalance the energy balance equation with diet only.

Practical Illustration

Why do short periods of caloric restriction often encourage dieters, but usually disappoint in the long run?

Results Not Always Predictable

List two reasons why the mathematics of weight loss are not always accurate for actual weight lost.

1.

2.

Set-Point Theory: A Case Against Dieting

Summarize the main premise of the set-point theory for body weight regulation.

Resting Metabolism Lowered

Describe what usually happens to the basal metabolic rate when weight loss is attempted through only dietary restriction.

Weight Cycling: Going No Place Fast

Define weight cycling; indicate its effect on chances for long-term weight control?

   Weight cycling:

   Chances for success:

How to Select a Diet Plan

Give three guidelines for establishing a prudent calorie-counting approach to weight loss. (Hint: Refer to Tables 19.1 and 19.2 in your textbook.)

1.

2.

3.

Well Balanced But Less of It

Give the composition of weight loss in the initial and later stages of a weight loss program?

Initial stage:

Later stage:

Maintenance of Goal Weight

Masters of Weight Control

List two important strategies that "masters of weight control" use for successful weight loss.

1.

2.

What is the preferred activity for men and women who succeed at weight loss?

Men:

Women:

EXERCISING TO TIP THE ENERGY BALANCE EQUATION

Explain how Americans now eat 5 to 10% fewer calories than they did 20 years ago, but they weigh an average 2.3 kg more.

Not Simply a Problem of Gluttony

State the two arguments that oppose the use of exercise as a means for weight control.

1.

2.

Pattern Also Holds for Children

What factor can contribute to the increase in obesity among children.

Increasing Energy Output

Present two arguments against using exercise for unbalancing the energy balance equation towards weight loss.

1.

2.

Exercise Effects on Food Intake

Does an increase in daily physical activity always results in a corresponding increase in daily food intake?

Exercise Effects on Energy Expenditure

What is the main benefit of exercise in terms of energy expenditure and weight control?

   Example one:

   Example two:

Exercise Alters Body Composition

List four factors to consider when designing an exercise program for weight control.

1.   3.

2.   4.

Dose–Response Relationship

Explain why weight loss relates to the amount of exercise time.

Maximize Total Energy Expenditure

What combination of exercise intensity and duration optimizes maximal energy expenditure for weight loss?

Start Slowly

In addition to progressing too rapidly in an exercise program, list three other variables that negatively affect an individual's compliance to regular exercise for weight loss.

1.

2.

3.

DIET PLUS EXERCISE: THE IDEAL COMBINATION

Why does the best approach for weight loss combine exercise and diet?

Discuss the recommended prudent limit for weekly weight loss.

Setting a Target Time

Calculate the average daily caloric deficit required to achieve a 20 pound fat loss in 20 weeks? Distribute the creation of this deficit equally between diet and exercise.

Total daily deficit:

Deficit due to diet:

Deficit due to exercise:

General Observations on Weight Loss

Write four statements that summarize the research on various approaches to weight loss.

1.

2.

3.

4.

Spot Reduction: Does it Work?

What effect does specific exercise at a particular body site have on fat loss at that site?

Where on the body does fat reduction occur most readily?

GAINING WEIGHT

In addition to increasing caloric intake, list two other important factors when attempting to gain weight?

Increase Lean, Not Fat

Discuss whether a person desiring to gain body mass should increase protein intake.

How Much Gain to Expect?

What increase in body mass can a man or woman expect after one year of heavy resistance training?

Man:

Woman:

Practice Quiz

MULTIPLE CHOICE

1. Which are true statements?
   a. The number of fat cells decreases with severe dietary restriction
   b. Body fat increases in previously nonobese adults occur by filling existing fat cells
   c. New adipocytes may also develop with further weight gain among the massively obese
   d. a and c
   e. b and c
2. Documented health risks of obesity include:
   a. Coronary heart disease
   b. Low blood pressure
   c. Hyperthyroidism
   d. All of the above
   e. None of the above

3. Limited data indicate that fat cell number in humans probably increases significantly:
   a. During the first trimester of pregnancy
   b. During the last trimester of pregnancy
   c. During the first year of life
   d. b and c
   e. All of the above
4. Severe dieting offers an ineffective approach to weight loss because?
   a. It increases water retention
   b. More lean tissue is lost compared with the same caloric deficit created by diet plus exercise
   c. The obese do not eat excessively so true gluttony is often not the problem
   d. b and c
   e. All of the above

5. Burning 10 kcal each minute burns the calories in one pound of body fat in about:
   a. 500 minutes
   b. 6 hours
   c. 3 hours
   d. 200 minutes
   e. None of the above

6. Enzyme that affects fat distribution pattern:
   a. Hexokinase
   b. Lipoprotein lipase
   c. Dehydrogenase
   d. Acetyl-CoA
   e. None of the above

7. Type of obesity that poses greatest health risk:
   a. Abdominal obesity
   b. Pear-type obesity
   c. Gynoid-type obesity
   d. a and b
   e. a and c

8. This many extra kcal equals approximately one pound of stored body fat:
   a. 2500
   b. 3000
   c. 3500
   d. 4000
   e. 6000

9. Even when obese people lose weight, achieving permanent fat loss remains difficult because?
   a. Adipocytes continue to hypertrophy
   b. Adipocytes increase their level of lipoprotein lipase
   c. Existing adipocytes increase by hyperplasia
   d. Adipocytes increase sensitivity to receptor sites
   e. Hyperplasia increases the blood supply to new adipocytes

10. The limit of weekly weight loss recommended for long term success is _____ pounds.
    a. < 2
    b. 2 to 3
    c. 3 to 4
    d. 4 to 5
    e. 5 to 6

TRUE/ FALSE

1. _____ The average adult gains weight during middle age despite a progressive decrease in food intake.

2. _____ Research indicates no genetic link to susceptibility to becoming obese.

3. _____ Diets that restrict water intake cause a greater weight loss, but the extra weight lost is mostly water.

4. _____ The calorie-expending effects of exercise are cumulative.

5. _____ Combining exercise with a low-calorie diet helps conserve the body's lean tissue during weight loss.

6. _____ When obesity begins in childhood, the chances for adult obesity becomes three times greater compared to normal weight children.

7. _____ The number of fat cells actually decreases with weight loss.

8. _____ All individuals have the same physiological "set-point."

9. _____ Localized exercise of a specific body area selectively increases fat loss from that area.

10. _____ Resistance training plus proper diet increases muscle mass and strength.

CROSSWORD PUZZLE

ACROSS

CHAPTER 19

2. _____ to hip ratio is important health indicator
4. university researcher (name starts with letter H) who pioneered studies of fat cellularity
9. repetitive bouts of dieting each followed by subsequent weight gain
12. fat storing enzyme
16. NIH views even low levels of excess body fat to be a health _____
18. male pattern obesity
19. excellent exercise mode for weight control
22. sampling of small amounts of fat from subcutaneous depots
24. word describing distribution of fat over the body
26. most adult weight gain due to this change in adipocytes

DOWN

1. synonym for lipid
3. localized fat loss with exercise
4. significant increase in number of adipocytes
5. obesity often begins in this stage of life
6. upper limit for recommended weekly pounds of weight loss
7. patterning where fat accumulates in lower body regions
8. % body fat is an important _____ for obesity
10. another word for obesity
11. 3500 kcal in one lb of _____ fat
13. opposite of gain or increase
14. when part of a weight loss program, more of the weight lost is fat
15. synonym for fat tissue
17. theory that body physiologically tries to maintain a particular level of body fat
20. not necessary for adults to _____ weight as they grow older
21. excessive amounts of body fat
23. _____ cycling is another term for yo-yo dieting
24. LBM is _____ when weight loss is accompanied by exercise (word begins with letter p)
25. area of body where fat may be more easily lost

Exercise, Aging, and Cardiovascular Health

1. Adrenopause

2. Andropause

3. Angina pectoris

4. Apoprotein

5. Atherosclerosis

6. Borderline high blood pressure

7. Cardiac arrest

8. CHD risk factors

9. Coronary heart disease

10. DHEA

11. Diastolic hypertension

12. Harvard alumni study

13. Heart attack

14. High-density lipoproteins

15. Homocysteine

16. Hyperlipidemia

17. Hyperlipoproteinemia

18. Hypothalamic-pituitary-gonadal axis

19. Ischemia

20. Low-density lipoproteins

21. Menopause

22. Myocardial infarction

23. Primary CHD risk factors

24. Somatopause

25. Successful aging

26. Surgeon General's Report on Physical Activity and Health

27. Systolic hypertension

28. Thrombus

29. Treatable risk factors

30. Type A personality

31. Type B personality

Study Questions

List the three objectives of the Surgeon General's Report on Physical Activity.

1.

2.

3.

List three of the Report's major conclusions.

1.

2.

3.

Physical Activity and Specific Health Concerns

List five beneficial effects of regular physical activity.

1.   4.

2.   5.

3.

SAFETY OF EXERCISING

What is the most prevalent injury caused by exercise.

THE NEW GERONTOLOGY

How long can one expect to live healthfully?

CONCEPT OF SUCCESSFUL AGING

Briefly explain the basic concept of "successful aging."

AGING AND BODILY FUNCTION

Draw a generalized curve that relates the level of body functions to age. (Hint: Refer to Fig. 20.1 in your textbook.)

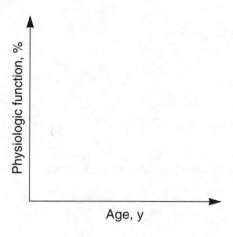

Aging and Muscular Strength

How much loss of muscular strength occurs by age 70?

Decrease in Muscle Mass

Reduced muscle mass triggers what age-associated decrease in physiologic function?

Muscle Trainability Among the Elderly

How much can the elderly increase their strength with a proper resistance training program? (HINT: Refer to Fig. 20.2 in your textbook.)

Aging and Joint Flexibility

Describe the change in joint flexibility with aging.

Endocrine Changes with Aging

List three "hormonal systems" most affected by aging.

1.

2.

3.

Aging and Nervous System Function
Describe the general effect of aging on the nervous system.

Aging and Pulmonary Function
Describe how aging affects pulmonary function.

Aging and Cardiovascular Function
Maximal Oxygen Uptake

How does aging affect the cardiovascular system?

Give the estimated maximum heart rate for a 59-year old person. Show your calculations.

Give the rate of decline in $\dot{V}O_{2max}$ with aging.

List two reasons that $\dot{V}O_{2max}$ declines with age.

1.

2.

Aging Response to Exercise Training

How much can a middle age person expect to increase $\dot{V}O_{2max}$ with proper training?

Aging and Endurance Performance

What is the percent difference in endurance performance between a 30-year-old and an 80-year-old marathoner? (Hint: Refer to Fig. 20.5 in your textbook.)

Other Age-Related Variables

Give three reasons for age-related decrements in HR_{max}.

1.

2.

3.

Aging and Body Composition

How much weight does the average 20-year-old male gain by age 60?

REGULAR EXERCISE: A FOUNTAIN OF YOUTH?

List five health benefits of regular physical activity.

1.   4.

2.   5.

3.

Does Exercise Improve Health and Extend Life?

Does exercise participation in youth ensure significant longevity?

Enhanced Quality to a Longer Life: A Study of Harvard Alumni

Describe major findings of the Harvard Alumni study.

Improved Fitness: A Little Goes a Long Way

Explain major findings of the experiment described in Figure 20.10 of your text.

CORONARY HEART DISEASE

What percentage of total deaths do diseases of the heart and blood vessels cause?

Women at Risk

Write four statements that summarize the heart disease risk for females.

1.   3.

2.   4.

A Life-Long Process

At what age can fatty streaks develop in the coronary arteries?

RISK FACTORS FOR CORONARY HEART DISEASE

List two major modifiable and fixed coronary heart disease risk factors.

    Modifiable risk factors:
      1.

      2.

    Fixed risk factors:
      1.

      2.

Age, Sex, and Heredity

Why do women possess a "gender advantage" for avoiding heart disease?

Blood Lipid Abnormalities

Cholesterol and Triglycerides

List desirable cholesterol and triglyceride levels for adults.

    Cholesterol:

    Triglyceride:

Importance of Cholesterol Forms

List four different lipoproteins.

1.   3.

2.   4.

List four most prevalent CHD risk factors in children and adolescents.

1.   3.

2.   4.

Homocysteine and Coronary Heart Disease

Describe the proposed effect of homocysteine on cholesterol.

List three vitamins that facilitate conversion of homocysteine to other nondamaging amino acids.

1.

2.

3.

Physical Activity and Coronary Heart Disease

How much greater is heart disease risk for a sedentary person compared with a physically active person?

Hypertension

Indicate the lower borderline limit for the classification of high blood pressure.

   Systolic:

   Diastolic:

List four severe medical complications caused by uncorrected chronic hypertension.

1.   3.

2.   4.

Exercise and Hypertension

Describe the effects of regular exercise on hypertension.

Cigarette Smoking

List three facts about cigarette smoking and the risk of developing CHD.

1.

2.

3.

Obesity

Discuss how obesity contributes as a multiple CHD risk factor.

Personality and Behavior Patterns

Describe personality characteristics of individuals who exhibit Type A and Type B behavior.

Type A:

Type B:

Physical Inactivity

List eight benefits of regular exercise for CHD prevention.

1.   5.

2.   6.

3.   7.

4.   8.

Interaction Among Risk Factors

Discuss the interaction among the primary CHD risk factors.

BEHAVIORAL CHANGES IMPROVE OVERALL HEALTH PROFILE

List the three most important health-saving tactics that reduce risk or bolster resistance to cancer, heart attack, stroke, and diabetes.

1.

2.

3.

Practice Quiz

MULTIPLE CHOICE

1. Regular physical activity:
   a. Lowers blood pressure
   b. Lowers blood cholesterol
   c. Normalizes blood clotting mechanisms
   d. Improves myocardial blood supply
   e. All of the above

2. Declines most with increasing age:
   a. Nerve conduction velocity
   b. Joint flexibility
   c. Lean body mass
   d. Liver and kidney function
   e. Reaction time

3. Reduced muscle mass with aging:
   a. Explains age-related decrease in skinfolds
   b. Explains age-related strength decrease
   c. Reflects increase in total muscle protein
   d. Reflects increase in total muscle fiber number
   e. All of the above

4. Formula for predicting maximum heart rate:
   a. $HR_{max} = 200 - age$
   b. $HR_{max} = 220 - age$
   c. $HR_{max} = age + 200$
   d. $HR_{max} = age + 220$
   e. None of the above

5. Percentage of total U.S. deaths caused by diseases of the heart and blood vessels:
   a. 25%
   b. 50%
   c. 65%
   d. 75%
   e. none of the above

6. True concerning CHD in men and women:
   a. Men more likely than women to have heart attacks during middle age
   b. Women are more likely than men to have a second heart attack
   c. Women survive bypass surgery at higher rates than men
   d. High blood sugar affects men more than women
   e. a and b

7. Desirable cholesterol level:
   a. ≤ 100 mg·dL^{-1}
   b. ≤ 200 mg·dL^{-1}
   c. ≤ 300 mg·dL^{-1}
   d. ≤ 400 mg·dL^{-1}
   e. None of the above

8. Regular exercise does not:
   a. Improve myocardial circulation and metabolism
   b. Enhance contractile properties of the myocardium
   c. Normalize the blood lipid profile
   d. Increase the levels of estrogen in the blood
   e. b and d

9. Include stress testing in a medical evaluation to:
   a. Screen for "silent" coronary disease
   b. Reproduce and access exercise-related chest symptoms
   c. Detect abnormal blood pressure response
   d. Monitor responses to various therapeutic interventions
   e. All of the above

10. Not an exercise-related indicator of coronary heart disease:
   a. Average maximum oxygen uptake
   b. Angina pectoris
   c. ECG disorders
   d. Extreme changes in blood pressure
   e. a and d

TRUE / FALSE

1. _____ The most prevalent exercise complication is sudden death during exercise.

2. _____ Children have a lower exercise economy during weight-bearing exercise than adults.

3. _____ No additional health or longevity benefits occur beyond weekly exercise of 3500 kcal.

4. _____ Increased coronary heart disease for women in later years relates to a loss of estrogen after menopause.

5. _____ Low density lipoproteins transport cholesterol to the liver where it is metabolized and excreted in the bile.

6. _____ High density lipoprotein levels increase with regular aerobic exercise.

7. _____ Diabetes, often called the "silent killer," generally progresses unnoticed for decades.

8. _____ For adults above 35 years of age, a medical evaluation is advised before any major increase in physical activity.

9. _____ A depressed S-T segment of an ECG usually indicates coronary heart disease.

10. _____ Cardiac patients generally do not improve their functional capacity to the same extent as a healthy individual of the same age.

CROSSWORD PUZZLE

ACROSS

CHAPTER 20

1. word describing distribution of fat over the body
5. % body fat is an important _____ for obesity
6. synonym for fat tissue
9. university researcher (name starts with letter H) who pioneered studies of fat cellularity
10. when part of a weight loss program, more of the weight lost is fat
14. fat storing enzyme
19. 3500 kcal in one lb of _____ fat
21. theory that body physiologically tries to maintain a particular level of body fat
22. NIH views even low levels of excess body fat to be a health _____
25. describes relationship between quantity of exercise and amount of weight loss
26. excellent exercise mode for weight control
27. male pattern obesity
28. another word for obesity

DOWN

2. area of body where fat may be more easily lost
3. patterning where fat accumulates in lower body regions
4. significant increase in number of adipocytes
7. opposite of gain or increase
8. not necessary for adults to _____ weight as they grow older
9. most adult weight gain due to this change in adipocytes
11. obesity often begins in this stage of life
12. sampling of small amounts of fat from subcutaneous depots
13. _____ to hip ratio is important health indicator
15. LBM is _____ when weight loss is accompanied by exercise (word begins with letter p)
16. upper limit for recommended weekly pounds of weight loss
17. localized fat loss with exercise
18. synonym for lipid
20. repetitive bouts of dieting each followed by subsequent weight gain
23. excessive amounts of body fat
24. _____ cycling is another term for yo-yo dieting

Clinical Exercise Physiology for Health-Related Professionals

1. ACSM

2. Aneurysm

3. Angina pectoris

4. Arrhythmias

5. Cardiac catheterization

6. Chronic bronchitis

7. Clinical exercise physiologist

8. Congestive heart failure

9. COPD

10. Coronary angiography

11. Coronary occlusion

17. Echocardiography

12. CPK

18. Ectopic

13. CT scanning

19. Emphysema

14. Cystic fibrosis

20. Endocarditis

15. Disability

21. GXT

16. Dyspnea

22. Handicap

23. Heart auscultation

24. LDH

25. Mital valve

26. MRI

27. Myocardial infarction

28. Oncology

29. Pericarditis

30. Prolapse

31. Regurgitation

32. RLD

33. SGOT

34. Sinus bradycardia

35. Sinus tachycardia

38. Thallium imaging

36. Stenosis

39. Ventriculography

37. Stress test sensitivity

40. Walk-through angina

Study Questions

THE EXERCISE PHYSIOLOGIST/HEALTH-FITNESS PROFESSIONAL IN THE CLINICAL SETTING

Describe the primary focus of the clinical exercise physiologist.

SPORTS MEDICINE AND EXERCISE PHYSIOLOGY: A VITAL LINK

List two activities of the clinical exercise physiologist in concert with other members of the health care team.

1.

2.

TRAINING AND CERTIFICATION BY PROFESSIONAL ORGANIZATIONS

Give the main focus of the health-fitness professional in relation to the U.S. Department of Health and Human Service's health objectives for the nation.

ACSM Qualifications and Certifications

Give two general objectives for an exercise leader in exercise physiology.

1.

2.

Health and Fitness Track

EXERCISE LEADER

HEALTH/FITNESS INSTRUCTOR

HEALTH/FITNESS DIRECTOR

List two required competencies of an ACSM certified exercise leader, health fitness instructor, and health fitness director.

Exercise leader

1.

2.

Health fitness instructor

1.

2.

Health fitness director

1.

2.

ACSM Clinical Track

The ACSM clinical track trains personnel for leadership in what areas?

EXERCISE TEST TECHNOLOGIST

PREVENTIVE/REHABILITATIVE EXERCISE SPECIALIST

PREVENTIVE/REHABILITATIVE PROGRAM DIRECTOR

List two required competencies of an ACSM certified exercise test technologist, preventive/rehabilitative exercise specialist and preventive/rehabilitative program director.

Exercise test technologist

1.

2.

Preventive/Rehabilitative exercise specialist

1.

2.

Preventive/rehabilitative program director

1.

2.

EXERCISE PROGRAMS FOR SPECIAL POPULATIONS
ONCOLOGY

List the three most prevalent forms of cancer.

1.

2.

3.

Describe the overall goal of the exercise specialist in cancer treatment.

Exercise Effects on Cancer

State two hypotheses about how exercise affects cancer.

1.

2.

Exercise Prescription and Cancer

Describe a prudent approach to exercising for cancer patients.

Does interval or continuous exercise provide the best initial approach for the cancer patient?

Breast Cancer and Exercise

List three primary risk factors for breast cancer.

1.

2.

3.

List two areas in which regular aerobic exercise benefits breast cancer patients.

1.

2.

CARDIOVASCULAR DISEASES

List three heart disease categories leading to functional disabilities. (Hint: Refer to Table 21.4 in your textbook.)

1.

2.

3.

List four different terms that generally refer to myocardial degenerative disease.

1.

2.

3.

4.

Diseases of the Myocardium

Outline five steps in CHD pathogenesis:

1.   4.

2.   5.

3.

ANGINA PECTORIS

List three symptoms of angina pectoris.

1.

2.

3.

Myocardial Infarction

Describe a major symptom of an acute MI.

Pericarditis

List four usual symptoms of pericarditis.

1.   3.

2.   4.

Congestive Heart Failure

List two common symptoms of CHF.

1.

2.

Aneurysm

Describe common symptoms of an aneurysm.

Heart Valve Diseases

List three heart valve abnormalities.

1.

2.

3.

Endocarditis

Describe symptoms of endocarditis.

Congenital Malformations

Congenital malformations of the heart occur about _____ times in every 100 births.

Mitral Valve Prolapse

Give the prevalence of floppy valve syndrome?

Cardiac Nervous System Diseases

List three diseases of the heart's electrical conduction system.

1.

2.

3.

CARDIAC DISEASE ASSESSMENT

List two typical symptoms in the differential diagnosis of chest pain.

1.

2.

Patient Medical History

List two symptoms of chest pain and their possible causes that can be obtained from a medical history.

    <u>Symptoms</u>   <u>Possible causes</u>

1.

2.

Physical Examination

List the four vital signs.

1.   3.

2.   4.

Heart Auscultation

List two heart abnormalities readily uncovered by heart auscultation.

1.

2.

Laboratory Tests

List five laboratory tests for CHD assessment.

1.   4.

2.   5.

3.

Noninvasive Physiological Tests

List two common noninvasive tests for cardiac dysfunction.

1.

2.

Echocardiography

Describe a major advantage of echocardiography?

Graded Exercise Stress Test (GXT)

State two main purposes of performing a GXT in the clinical setting.

1.

2.

What is the most common GXT mode?

WHY STRESS TEST?

Give four reasons to justify stress testing.

1.   3.

2.   4.

WHO SHOULD BE STRESS-TESTED?

What groups of people need not be stress tested prior to initiating an exercise program?

INFORMED CONSENT

List two major components of a properly written informed consent.

1.

2.

CONTRAINDICATIONS TO STRESS TESTING

ABSOLUTE CONTRAINDICATIONS

List eight medical conditions that require medical supervision during stress testing.

1.   5.

2.   6.

3.   7.

4.   8.

RELATIVE CONTRAINDICATIONS

List seven medical conditions that require that medical personal are in the area of the stress test.

1.   5.

2.   6.

3.   7

4.

Field Tests of Cardiorespiratory Fitness

List two common field tests to estimate cardiorespiratory fitness.

1.

2.

Maximal Versus Submaximal Tests

Define "symptom limited GXT?"

State the main advantage and main disadvantage of a GXT_{max} test?

 Advantage:

 Disadvantage:

List six factors that influence physiologic response to submaximum or maximum exercise.

1.   4.

2.   5.

3.   6.

STRESS TEST PROTOCOLS

Treadmill Tests

List three treadmill stress test protocols, and give one advantages and disadvantage of each.

Protocol	Advantage	Disadvantage
1.		
2.		
3.		

Bicycle Ergometer Tests

List three advantages of bicycle ergometry for stress testing.

1.

2.

3.

Arm-Crank Ergometer Tests

How much lower is the $\dot{V}O_{2max}$ and HR_{max} response in arm-crank exercise compared with exercise on a treadmill or cycle ergometer.

 Treadmill:

 Bicycle ergometer:

MAXIMAL TREADMILL, CYCLE ERGOMETER, AND SWIMMING TESTS

Which would result in the highest, mid, and lowest $\dot{V}O_{2max}$.

SAFETY OF STRESS TESTING

State the approximate death risk per 10,000 stress tests.

Stress Test Outcomes

Name and explain the four possible outcomes of a stress test.

1.   3.

2.   4.

What is the most prevalent stress test outcome?

EXERCISE-INDUCED INDICATORS OF CHD

Angina Pectoris

Angina pains indicate what condition for the myocardium?

ECG Disorders

List three conditions that ECG electrical disorders may indicate.

1.

2.

3.

List three types of S-T segment disorders.

1.

2.

3.

CARDIAC RHYTHM ABNORMALITIES

Arrhythmia's with exercise frequently appear as _____.

Individuals who experience frequent PVC's during exercise have a high risk of _____ _____ from ventricular fibrillation.

Other Indices of CHD

HYPERTENSIVE RESPONSE

Indicate the normal upper limit for systolic and diastolic blood pressure during a GXT.

Systolic:

Diastolic:

HYPOTENSIVE RESPONSE

What condition may indicate failure of blood pressure to increase with exercise?

HEART RATE RESPONSE

_____ indicates a rapid heart rate.

_____ indicates a slow heart rate.

INVASIVE PHYSIOLOGIC TESTS

List two conditions that invasive physiological testing can diagnose that cannot be evaluated through noninvasive procedures.

1.

2.

Radionucleotide Studies

List two types of radionuleotide procedures.

1.

2.

Cardiac Catheterization

Is cardiac catheterization considered an invasive or noninvasive procedure?

Coronary Angiography

Angiography serves as the criterion "gold standard" for viewing coronary _____ _____.

PATIENT CLASSIFICATION FOR CARDIAC REHABILITATION

State the usual first sign of cardiac disease?

List three symptoms that accompany cardiac disease progression.

1.

2.

3.

PHASES OF CARDIAC REHABILITATION

List three important aspects for a successful cardiac rehabilitation program.

1.

2.

3.

List three phases of a cardiac rehabilitation program. (Hint: Refer to Table 21.14 in your textbook.)

1.

2.

3.

Describe two functional changes resulting from a properly prescribed exercise program that reduce myocardial demands during exercise.

1.

2.

EXERCISE PRESCRIPTION

Guidelines

Describe the aspect of the exercise prescription most difficult to determine?

List two "systems" for establishing training intensity.

1.

2.

What does FITT represent?

_____

_____

_____

_____

Practical Illustration

In addition to an exercise prescription, list three additional components of cardiac rehabilitation.

1.

2.

3.

Beneficial Effects of Resistance Exercises

State the best type of resistance exercise for the CHD patient.

Describe two benefits of resistance training for the CHD patient.

1.

2.

THE REHABILITATION PROGRAM

List optimum exercise type, duration, intensity, and frequency for an effective preventive or rehabilitative exercise program.

Exercise type:

Duration:

Intensity:

Frequency:

Cardiac Rehabilitation Outcomes

List four outcomes from a cardiac rehabilitation program.

1.   3.

2.   4.

Exercise Training

Available research indicates _____ as the strongest outcome measure from a cardiac rehabilitation program.

Education, Counseling, and Behavioral Interventions

List three positive "nonmedical" outcome effects of a cardiac rehabilitation program.

1.

2.

3.

Supervision Level

What type of person can enroll in an unsupervised cardiac rehabilitation program?

Program Administration

List three issues that should concern cardiac rehabilitation program administrators.

1.

2.

3.

Give an example of a mission statement for a cardiac rehabilitation program.

List five types of personnel that work in a full-service cardiac rehabilitation program.

1.   4.

2.   5.

3.

CARDIAC MEDICATIONS

List six classifications of common cardiac drugs.

1.   4.

2.   5.

3.   6.

PULMONARY DISEASES

List two major classifications of pulmonary abnormalities.

1.

2.

Restrictive Lung Dysfunction

List two major characteristics of RLD.

1.

2.

Explain lung compliance.

List five major types of RLD. (Hint: Refer to Table 21.21 in your textbook.)

1.   4.

2.   5.

3.

Chronic Obstructive Pulmonary Disease

List three of the classic symptoms of COPD

1.

2.

3.

Chronic Bronchitis

Describe two symptoms and characteristics of chronic bronchitis.

1.

2.

Emphysema

Describe characteristics and give the cause of the "emphysemic look."

Characteristics:

Cause:

Cystic Fibrosis

State the objective diagnosis for CF.

PULMONARY ASSESSMENTS

List five major pulmonary assessment procedures.

1.   4.

2.   5.

3.

Pulmonary Rehabilitation and Exercise Prescription

List five goals of pulmonary rehabilitation.

1.   4.

2.   5.

3.

Describe the perceived dyspnea scale.

How does the exercise prescription differ between patients with moderate and severe lung disease?

PULMONARY MEDICATIONS

List three common pulmonary drugs and their actions.

1.

2.

3.

Practice Quiz

MULTIPLE CHOICE

1. The highest incidence of cancer affect:
   a. Breasts, lungs, bowl, and uterus
   b. Skin, lungs, pancreas, and kidney
   c. Skin, lungs, and intestines
   d. Breasts and skin
   e. Myocardium

2. Disability:
   a. Defined handicap
   b. Diminished functional capacity
   c. Physical performance limitation
   d. Particular disease
   e. None of the above

3. Regular exercise:
   a. No effect on lowering cancer risk
   b. No effect on lowering cancer risk but aids cancer rehabilitation
   c. Decreases cancer risk by increasing levels of anti-inflammatory cytokines
   d. Decreases cancer risk by increasing antioxidant tolerance
   e. c and d

4. Exercise as a therapy for breast cancer:
   a. Decreases levels of depression and anxiety
   b. Has little effect on psycho-social variables
   c. Decreases plasma cytokines which helps healing
   d. Benefits only older women
   e. a and d

5. Diseases of the myocardium:
   a. Angina pectoris, myocardial infarction, pericarditis, congestive heart failure, and aneurysms
   b. Stenosis, regurgitation, and prolapse
   c. Dysrhythmias and endocarditis
   d. Prolapse and endocarditis
   e. None of the above

6. Most angina pain occurs in the area of the _____ _____ and lasts for less than _____ minutes:
   a. Left shoulder and down inside of left arm; 5
   b. Right shoulder and head; 20
   c. Abdominal area; 5
   d. Abdominal area and legs; 20
   e. Back and hips; 5

7. A myocardial infarction:
   a. Increase in heart's H^+ ion concentration, which decreases contractility
   b. Decrease in pulmonary blood flow due to reduced alveolar PO_2
   c. Deprivation of blood (oxygen) to a portion of the myocardium
   d. Inflammation of the heart's pericardium
   e. Decreases in the hormones epinephrine and norepinephrine

8. Increased levels of _____ , _____ , and _____ can verify the presence of an acute MI.
   a. CP, SDH, SGOT
   b. LDH, PFK, CP
   c. ADP, SDH, PK
   d. CPK, LDH, SGOT
   e. None of the above

9. The most difficult yet important aspect of the exercise prescription:
   a. Exercise duration
   b. Patient's initial functional capacity
   c. Exercise intensity
   d. Exercise frequency
   e. Patient's MET capacity

10. Diagnosis of cystic fibrosis:
    a. Depressed ST segment of the ECG
    b. Positive sweat electrolyte (chloride)
    c. Decreased residual lung volume
    d. Decreased vital capacity during moderate exercise
    e. a and b

TRUE/FALSE

1. _____ For individuals whose occupations predominantly require arm work, exercise training should emphasize this musculature because of exercise specificity in the training response.

2. _____ Ventriculography provides an intracardiac X-ray after a radiopaque contrast medium enters the coronary blood vessels.

3. _____ Sinus bradycardia occurs normally in endurance athletes and young adults.

4. _____ Valvular stenosis refers to constriction or narrowing of the heart valves.

5. _____ Diagnosis of COPD requires pulmonary function testing.

6. _____ COPD is common among sedentary non-smokers who eat high-fat foods.

7. _____ Regular aerobic exercise provides benefits for cystic fibrosis patients.

8. _____ Regular aerobic exercise improves respiratory muscle function in pulmonary disease patients.

9. _____ A person with advanced emphysema often has a normal appearing chest on visual inspection.

10. _____ MRI scans are of limited use in diagnosing pulmonary disease.

CROSSWORD PUZZLE

ACROSS

1. this type of exercise physiologist works in a health related setting
7. inflammation of the heart's inner lining
10. balloon-like, blood-filled sac in arterial wall
11. term for heart attack (abbr.)
13. inability of myocardium to pump blood at a life sustaining rate (abbr.)
14. process of inserting flexible tube to administer drugs, extract fluid, or monitor pressure
17. technique using ultrasound to study the heart
24. condition making a person unable to perform tasks expected by society
26. gene for susceptibility to atherosclerosis (abbr.)
28. second leading cause of death in US
29. heart beat resulting from momentary loss of normal rhythm
30. enzyme elevated following MI (abbr.)

DOWN

2. aspect of exercise that is most difficult in formulating the exercise prescription
3. enzyme elevated following an MI (abbr.)
4. occurs when a heart valve does not close properly; also called regurgitation
5. slowing of the heart rate
6. blockage of blood flow to myocardim
8. chest pain from myocardial ischemia

9. congenital metabolic disorder affecting pancreatic and bile ducts and bronchi
12. progressive incremental exercise stress test (abbr.)
15. speeding up of the heart rate
16. one of the most common forms of cancer in caucasian women
18. sweat electrolyte used to diagnose cystic fibrosis
19. study of and treatment of cancer
20. labored breathing causing distress, especially shortness of breath
21. inflammation of the heart's outer lining
22. obstructive lung disease (abbr.)
23. common form of COPD
25. enzyme elevated following an MI (abbr.)
27. also called floppy valve syndrome (abbr.)

Section II

- Self-Assessment Tests

Physical Activity Readiness (PAR-Q)

Complete the following Physical Activity Readiness Questionnaire (The PAR-Q[a]) to get a general idea of whether you are physically ready to exercise.

PAR-Q is designed to help you help yourself. Many health benefits are associated with regular exercise, and the completion of PAR-Q is a sensible first step if you are planning to increase your physical activity. For most people, physical activity should not pose any problem or health hazard. PAR-Q has been designed to identify the small number of adults for whom physical activity might be inappropriate, or those who should have medical advice concerning the type of activity most suitable for them.

COMMON SENSE IS YOUR BEST GUIDE IN ANSWERING THESE QUESTIONS. READ EACH QUESTION CAREFULLY AND CHECK YES OR NO IF IT APPLIES TO YOU.

YES _____ NO _____ 1. Has your doctor ever said you have heart trouble?

YES _____ NO _____ 2. Do you frequently have pains in your heart and chest?

YES _____ NO _____ 3. Do you often feel faint or have spells of severe dizziness?

YES _____ NO _____ 4. Has a doctor ever said your blood pressure was too high?

YES _____ NO _____ 5. Has your doctor told you that you have a bone or joint problem that has been aggravated by exercise, or might be made worse with exercise?

YES _____ NO _____ 6. Is there a good physical reason not mentioned here why you should not follow an activity program even if you wanted to?

YES _____ NO _____ 7. Are you over age 65 and not accustomed to vigorous exercise?

IF YOU ANSWERED YES TO ONE OR MORE QUESTIONS

If you have not recently done so, consult with your personal physician by telephone or in person **BEFORE** increasing your physical activity and/or taking a fitness test. Show your doctor a copy of this quiz. After medical evaluation, seek advice from your physician as to your suitability for:

- Unrestricted physical activity, probably on a gradually increasing basis
- Restricted or supervised activity to meet your specific needs, at least on an initial basis. Check in your community for special programs or services

---

[a]PAR-Q was developed by the British Columbia Ministry of Health. Conceptualized and critiqued by the Multidisciplinary Advisory Board on Exercise (MABE). Reference: PAR-Q Validation Report, British Columbia Ministry of Health, May 1978.

IF YOU ANSWERED NO TO ALL QUESTIONS

If you answered PAR-Q accurately, you have reasonable assurance of your present suitability for:

- *A graduated exercise program:* a gradual increase in proper exercise promotes good fitness development while minimizing or eliminating discomfort
- *An exercise test:* simple tests of fitness (such as the Canadian Home Fitness Test) or more complex types may be undertaken if you so desire
- *Postpone exercising:* if you have a temporary minor illness, such as a common cold, postpone any exercise program

Revised Physical Activity Readiness Questionnaire (rPar-Q)

One limitation of the original PAR-Q was that a relatively large number (about 20%) of potential exercisers failed the test—and many of these exclusions were unnecessary as subsequent evaluation of these men and women showed that they were apparently healthy. Consequently, to reduce the number of unnecessary exclusions (false-positives) the revised PAR-Q (rPar-Q) was developed. This revision is a recommended method for determining the exercise readiness of apparently healthy middle-aged adults with no more than one major risk factor for coronary heart disease.

YES _____ NO _____ 1. Has a doctor said that you have a heart condition and recommended only medically supervised activity?

YES _____ NO _____ 2. Do you have chest pain brought on by physical activity?

YES _____ NO _____ 3. Have you developed chest pain in the past month?

YES _____ NO _____ 4. Do you tend to lose consciousness or fall over as a result of dizziness?

YES _____ NO _____ 5. Do you have a bone or joint that could be aggravated by the proposed physical activity?

YES _____ NO _____ 6. Has a doctor ever recommended medication for your blood pressure or a heart condition?

YES _____ NO _____ 7. Are you aware through your own experience, or a doctor's advice, of any other physical reason against your exercising without medical supervision

NOTE: Postpone testing if you have a temporary illness like a common cold, or are not feeling well.

IF YOU ANSWERED YES TO ONE OR MORE QUESTIONS

If you have not recently done so, consult with your personal physician by telephone or in person **BEFORE** increasing your physical activity and/or taking a fitness test. Show your doctor a copy of this quiz. After medical evaluation, seek advice from your physician as to your suitability for:

- Unrestricted physical activity, probably on a gradually increasing basis
- Restricted or supervised activity to meet your specific needs, at least on an initial basis. Check in your community for special programs or services

(From: Shephard, R. J., et al.: The Canadian Home Fitness Test: Update. Sports Med, 1:359, 1991.)

Eating Smart Assessment

(Source: American Cancer Society, Rev. 1989, pp. 2–5)

The *Eating Smart Assessment* gives a broad view of the diversity of the diet, especially its content of fat- and fiber-rich foods. A high rating means you're on the right track in terms of prudent nutrition to help fight the battle against heart disease and certain cancers.

OILS AND FATS: butter, margarine, shortening, mayonnaise, sour cream, lard, oil	POINTS
I always add these foods in cooking and/or at the table	_____ 0
I occasionally add these to foods in cooking and/or at the table	_____ 1
I rarely add these foods in cooking and/or at the table	_____ 2

DAIRY PRODUCTS: milk, yogurt, cheese, ice cream	POINTS
I drink whole milk	_____ 0
I drink 1 or 2% fat-free milk	_____ 1
I seldom eat frozen desserts or ice cream	_____ 2
I eat ice cream almost every day	_____ 0
Instead of ice cream, I eat ice milk, low-fat frozen yogurt and sherbet	_____ 1
I eat only fruit ices, seldom eat frozen dairy desserts	_____ 2
I eat mostly high-fat cheese (jack, cheddar, colby, Swiss, cream)	_____ 0
I eat both low- and high-fat cheeses	_____ 1
I eat mostly low-fat cheeses (2% cottage, skim milk, mozzarella)	_____ 2

SNACKS: potato/corn chips, nuts, buttered popcorn, candy bars	POINTS
I eat these every day	_____ 0
I eat some of these occasionally	_____ 1
I seldom or never eat these snacks	_____ 2

BAKED GOODS: pies, cakes, cookies, sweet rolls, doughnuts	POINTS
I eat them 5 or more times a week	_____ 0
I eat them 2–4 times a week	_____ 1
I seldom eat baked goods or eat only low-fat baked goods	_____ 2

POULTRY AND FISH: (If you do not eat meat, fish, or poultry, give yourself 2 points)	POINTS
I rarely eat these foods	_____ 0
I eat them 1–2 times a week	_____ 1
I eat them 3 or more times a week	_____ 2

LOW-FAT MEATS: extra lean hamburger, round steak, pork loin, roast, tenderloin, chuck roast. (If you do not eat meat, fish, or poultry, give yourself 2 points)	POINTS
I rarely eat these foods	_____ 0
I occasionally eat these foods	_____ 1
I eat mostly fat-trimmed red meats	_____ 2

HIGH-FAT MEATS: luncheon meats, bacon, hot dogs, sausage, steak, regular and lean ground beef. (If you do not eat meat, fish, or poultry, give yourself 2 points)	POINTS
I eat these every day	_____ 0
I occasionally eat these foods	_____ 1
I rarely eat these foods	_____ 2

CURED AND SMOKED MEAT AND FISH: luncheon meats, hot dogs, bacon, ham and other smoked or pickled meats and fish. (If you do not eat meat, fish, or poultry, give yourself 2 points)	POINTS
I eat these foods 4 or more times a week	_____ 0
I eat some of these foods 1–3 times a week	_____ 1
I seldom eat these foods	_____ 2

LEGUMES: dried beans, peas (kidney, navy, lima, pinto, garbanzo, split-pea, lentil)	POINTS
I eat legumes less than once a week	_____ 0
I eat legumes 1–2 times a week	_____ 1
I eat legumes 3 or more times a week	_____ 2

WHOLE GRAINS AND CEREAL: whole grain breads, brown rice, pasta, grain cereals	POINTS
I seldom eat these foods	_____ 0
I eat these foods 1–2 times a day	_____ 1
I eat these foods 4 or more times daily	_____ 2

VITAMIN-C RICH FRUITS AND VEGETABLES: citrus fruits, juices, green peppers, berries	POINTS
I seldom eat these foods	_____ 0
I eat these foods 3–5 times a week	_____ 1
I eat these foods 1–2 times a day	_____ 2

DARK GREEN AND DEEP YELLOW FRUITS AND VEGETABLES: broccoli, greens, carrots, peaches (dark green and yellow fruits and vegetables contain beta carotene that your body turns into vitamin A. Vitamin A helps protect against certain types of cancer-causing substances)	POINTS
I seldom eat these foods	_____ 0
I eat these foods 1–2 times a week	_____ 1
I eat these foods 3–4 times a week	_____ 2

VEGETABLES OF THE CABBAGE FAMILY: broccoli, cabbage, brussels sprouts, cauliflower	POINTS
I seldom eat these foods	_____ 0
I eat these foods 1–2 times a week	_____ 1
I eat these foods 3–4 times a week	_____ 2

ALCOHOL:	POINTS
I drink more than 2 oz daily	_____ 0
I drink every week, but not daily	_____ 1
I occasionally or never drink alcohol	_____ 2

YOUR BODY WEIGHT:	POINTS

I am more than 20 lb over my ideal weight _____ 0

I am 10–20 lb over my ideal weight _____ 1

I am within 10 lb of my ideal weight _____ 2

ADD UP YOUR TOTAL POINTS HERE   _____ TOTAL POINTS

YOUR EATING SMART RATING

0–12 Points: A Warning Signal

Your diet is too high in fat and too low in fiber-rich foods. Assess your eating habits to see where you could make improvements.

13–17 Points: Not Bad

You still have a way to go. Review your quiz to identify those areas in which you rate poorly, then make the necessary adjustments.

18–36 Points: Good For You, You're Eating Smart

You should feel very good about yourself. You have been careful to limit your fats and eat a varied diet. Keep up the good habits and continue to look for ways to improve.

C

Your Weight and Heart Disease I.Q.

(From the National Heart, Lung, and Blood Institute, National Institutes of Health)

Answer the following statements as either <u>true</u> or <u>false</u>. The statements test your knowledge about excess weight and heart disease. The correct answers can be found on the next page.

T
F
   1. Being overweight puts you at risk for heart disease.

T
F
   2. If you are overweight, losing weight helps lower your high blood cholesterol and high blood pressure.

T
F
   3. Quitting smoking is healthy, but it commonly leads to excessive weight gain that increases your risk for heart disease.

T
F
   4. An overweight person with high blood pressure should pay more attention to a low-sodium diet than to weight reduction

T
F
   5. A reduced intake of sodium or salt does not always lower high blood pressure to normal.

T
F
   6. The best way to lose weight is to eat fewer calories and to increase exercise.

T
F
   7. Skipping meals is a good way to cut down on calories.

T
F
   8. Foods high in complex carbohydrates (starch and fiber) are good choices when you are trying to lose weight.

T
F
   9. The single most important change most people can make to lose weight is to avoid sugar.

T
F
  10. Polyunsaturated fat has the same number of calories as saturated fat.

T
F
  11. Overweight children are very likely to become overweight adults.

Answers to Your Weight and Heart Disease I.Q. test

1. **TRUE.** Being overweight increases your risk for high blood cholesterol and high blood pressure. Even if you do not have high cholesterol or high blood pressure, being overweight may increase your risk for heart disease. Where you carry extra weight also may affect your risk. Weight carried at your waist or trunk seems to be associated with an increased risk for heart disease in many people.

2. **TRUE.** If you are overweight, even moderate reductions in weight, such as 5 to 10%, can produce substantial reduction in blood pressure. Moderate weight loss may also reduce your LDL-cholesterol and increase your HDL-cholesterol.

3. **FALSE.** The average weight gain after quitting smoking is 5 lb. The proportion of ex-smokers who gain large amounts of weight (>20 lb) is relatively small. Even if you gain weight when you stop smoking, you should change your eating and exercise habits to lose weight rather than starting to smoke again. Smokers who quit smoking decrease their risk for heart disease by more than 50%!

4. **FALSE.** If you are overweight, weight loss may reduce your blood pressure even if you don't reduce the amount of sodium consumed. Weight loss is recommended for all overweight people who have high blood pressure. Even if weight loss does not reduce blood pressure to normal, it may help reduce blood pressure medications.

5. **TRUE.** While a high sodium or salt intake plays a key role in maintaining high blood pressure, there is no easy way to determine who benefits from consuming less sodium and salt. Also, a high sodium or salt intake may limit how well certain high blood pressure medications work.

6. **TRUE.** Eating fewer calories and exercising more is the best way to lose weight and keep it off. A steady weight loss of 1–2 lb per week is safe, and the weight is more likely to stay off. Losing weight, if you are overweight, may also help reduce blood pressure and raise HDL-cholesterol levels.

7. **FALSE.** To cut calories, some people regularly skip meals, snacks, or caloric drinks. If you do this, your body thinks that it is starving even if your intake is not reduced. Your body will try to save energy by slowing its metabolism. This makes losing weight harder and may even add body fat.

8. **TRUE.** Contrary to popular belief, foods high in complex carbohydrates are lower in calories than foods high in fat. In addition, complex carbohydrates are good sources of vitamins, minerals, and fiber.

9. **FALSE.** Sugar has not been found to cause obesity; however, many foods high in sugar are also high in fat. Fat has more than twice the calories as the same amount of protein or carbohydrate.

10. **TRUE.** All fats—polyunsaturated, monounsaturated, and saturated—have the same number of calories. All calories count regardless of their source. Fats are the richest sources of calories; thus eating less total fat will help reduce the total number of calories consumed daily.

11. **FALSE.** Obesity in childhood does increase the likelihood of adult obesity, but most overweight children will not become obese. Several factors influence whether an overweight child becomes an overweight adult. These include age of onset of overweight, degree of childhood and family history of overweight, and dietary and activity habits.

YOUR SCORE

10–11 CORRECT: Congratulations! You know a lot about weight and heart disease. Share this information with your family and friends.

8–9 CORRECT: Very good.

<8 CORRECT: Go over the answers and try to learn more about weight and heart disease. Refer to Chapter 29 and 30 of your textbook.

D

Healthy Heart I.Q.

(From the National Heart, Lung, and Blood Institute, National Institutes of Health)

Answer the following statements either **true** or **false**. The statements test your knowledge about heart disease and its risk factors. The correct answers are on the next page.

T
F   1. The risk factors for heart disease that you *can do something about are:* high blood pressure, high blood cholesterol, smoking, obesity, and physical inactivity.

T
F   2. A stroke is often the first symptom of high blood pressure, and a heart attack is often the first symptom of high blood cholesterol.

T
F   3. A blood pressure greater than or equal to 140/90 mm Hg is generally considered high.

T
F   4. High blood pressure affects the same number of blacks as it does whites.

T
F   5. The best ways to treat and control high blood pressure are to control your weight, exercise, eat less salt, restrict alcohol intake, and take your prescribed high blood pressure medicine.

T
F   6. A blood cholesterol level of 240 mg·dL^{-1} is desirable for adults.

T
F   7. The most effective dietary way to lower your blood cholesterol is to eat foods low in cholesterol.

T
F   8. Lowering blood cholesterol levels can help people who have already had a heart attack.

T
F   9. Only children from families at high risk of heart disease need to check their blood cholesterol levels.

T
F   10. Smoking is a major risk factor for four of the five leading causes of death, including heart attack, stroke, cancer, and lung disease such as emphysema and bronchitis.

T
F   11. If you have had a heart attack, quitting smoking can help reduce your chances of having a second attack.

T
F   12. Someone who has smoked for 30 to 40 years probably will not be able to quit smoking.

T
F

13. Heart disease is the leading killer of men and women in the United States.

ANSWERS TO YOUR HEALTHY HEART I.Q. TEST

1. **TRUE.** High blood pressure, smoking, and high cholesterol are the three most important risk factors. On average, each factor doubles your chance of developing heart disease. Thus, a person who has all three risk factors is 8 times more likely to develop heart disease than someone who has none. Obesity increases the likelihood of developing high blood cholesterol and high blood pressure. Increased activity decreases heart disease risk since exercisers are more likely to cut down or stop smoking.

2. **TRUE.** A person with high blood pressure or high blood cholesterol may feel fine and look great, and there are often no signs of anything wrong.

3. **TRUE.** A blood pressure of 140/90 mm Hg or greater is classified as high. If the lower number is 85 or greater, this is considered a high risk.

4. **FALSE.** High blood pressure is more common in blacks than whites, and affects 40 out of every 100 black adults compared to 25 out of every 100 white adults. With aging, high blood pressure is more severe among blacks than among whites.

5. **TRUE.** Lifestyle changes can help keep blood pressure levels normal even into advanced age, and are important in treating and preventing high blood pressure.

6. **FALSE.** A total blood cholesterol of under 200 mg·dL^{-1} is desirable and puts you at a lower risk for heart disease. A level of 200–239 mg·dL^{-1} is considered borderline high and usually increases your risk for heart disease. All adults 20 years of age or older should have their cholesterol checked at least every 5 years.

7. **FALSE.** Reducing cholesterol is important; however, eating foods low in saturated fat is the most effective dietary way to lower cholesterol levels along with eating less total fat and cholesterol.

8. **TRUE.** People who have had one heart attack are at much higher risk for a second attack. Reducing cholesterol can slow down (and in some cases even reverse) and reduce the chances of a second attack.

9. **TRUE.** Children from "high" risk families should have their cholesterol checked at least every other year.

10. **TRUE.** Heavy smokers are 2 to 4 times more likely to have a heart attack than non-smokers. The heart attack death rate among all smokers is 70% greater than non-smokers. The risk of dying of lung cancer is 22 times higher for male smokers than non-smokers, and 12 times higher for female smokers than female non-smokers. Eighty percent of all deaths from lung disease are due directly to smoking.

11. **TRUE.** One year after quitting, ex-smokers cut their extra risk by about half or more, and eventually the risk returns to normal in healthy ex-smokers.

12. **FALSE.** Older smokers are more likely to succeed at quitting smoking than younger smokers.

13. **TRUE.** Heart disease is the #1 killer in the U.S. Approximately 489,000 Americans died of heart disease in 1990, and approximately half of these deaths were women.

YOUR SCORE

12–13 CORRECT: Congratulations! You know a lot about heart disease. Share this information with your family and friends.

10–11 CORRECT: Very good.

<10: CORRECT: Go over the answers and try to learn more about weight and heart disease. Refer to Chapters 29 and 30 of your textbook.

E

Assessment of Energy and Nutrient Intake

Three-Day Dietary Survey

Your three-day dietary survey is a relatively simple yet accurate method to determine the nutritional quality and total calories of the food consumed each day. The secret is to keep a daily log of food intake for any three days that represent your normal eating pattern.

Experiments have shown that calculations of caloric intake made from records of daily food consumption are usually within 10% of the number of calories actually consumed. For example, suppose the caloric value of your daily food intake was directly measured in a bomb calorimeter and averaged 2130 kcal. If you kept a three-day dietary history and estimated your caloric intake, the daily value would likely be within 10% of the actual value, or between 1920 kcal and 2350 kcal. Thus, as long as you maintain a careful record, the degree of accuracy for daily determinations will be within acceptable limits.

Before recording your daily caloric intake over the three-day period, you should become familiar with "honest" calorie counting. This requires four items: a plastic ruler, a standard measuring cup, measuring spoons, and an inexpensive balance or weighing scale. You can purchase these items at most hardware stores. Second, you should familiarize yourself with the nutritional and caloric values of foods by consulting food labels and Section IV of this Study Guide. Listed under "Specialty and Fast-Food Items" in the food listings are the nutritional values for items sold at fast-food chain stores.

Measure or weigh each of the food items in your diet. This is the only reliable way to obtain an accurate estimate of the size of a food portion. If you elect to use Section IV of this Study Guide to estimate the nutritional and kcal values, you only need to weigh each food item. If you use a supplementary calorie-counting guide, this may require the measuring cup, spoons, and ruler.

Food Categories

Meat and Fish

Measure the portion of meat or fish by thickness, length, and width, or record weight on a scale.

Vegetables, Potatoes, Rice, Cereals, Salads

Measure the portion in a measuring cup or record weight on a scale.

Cream or Sugar Added to Coffee or Tea

Measure with measuring spoons before adding to the drink, or record weight on a scale.

Fluids and Bottled Drinks

Check the labels for volume or empty the container into a measuring cup. If you weigh the fluid, be sure to subtract the weight of the cup or glass. Sugar-free soft drinks usually have kcal values listed on their labels.

Cookies, Cakes, Pies

For these items, measure the diameter and thickness with a ruler, or weigh on a scale. Evaluate frosting or sauces separately.

Fruits

Cut in half before eating and measure the diameters, or weigh on a scale. For fruits that must be peeled or have rinds or cores, be sure the weight of the non-edible portion is subtracted from the total weight of the food. Do this for items such as oranges, apples, and bananas.

Jam, Salad Dressing, Catsup, Mayonnaise

Measure with a measuring spoon or weigh the portion on a scale.

Directions For Computing Your Three-Day Dietary Survey

Step 1. Prepare a table (similar to Table E.1) indicating the intake of food items during a day. Include the amount (g or oz); caloric value; and carbohydrate, lipid, and protein content; the minerals Ca and Fe; and vitamins C, B1 (thiamine), and B2 (riboflavin); fiber; and cholesterol.

Step 2. Be sure to list <u>each</u> food you consume for breakfast, lunch, dinner, between-meal eating, and snacks. Include food items that are used in preparing the meal (e.g., butter, oils, margarine, bread crumbs, egg coating, etc.).

Step 3. Weigh, measure or approximate the size of each portion of food that you eat. Record these values on your daily record chart (e.g., 3 oz of salad oil, 1/8 piece of 8" diameter apple pie, etc.).

Step 4. Record your daily calorie and nutrient intake on a chart similar to Table E.1, which was recorded for a 21-year-old college student. Record the daily totals for the caloric and nutrient headings on the "Daily and Average Daily Summary Chart" (Table E.2). When you've completed your three-day survey, compute the three-day total by adding up the values for days 1, 2, and 3; then divide by 3 to determine the daily average of each nutrient category.

Step 5. Using each of the average daily nutrient values, calculate the percent of the RDA consumed for that particular nutrient and graph your results as shown in Figure E.1. An example for calculating the percent of the RDA is shown in Table E.3, along with the specific RDA values for men and women.

Step 6. Be as accurate and honest as possible. <u>Do not</u> include unusual or atypical days in your dietary survey (e.g., days that you are sick, special occasions such as birthdays, or eating out at restaurants unless that is normal for you).

Step 7. Remember that the protein RDA is equal to 0.8 g protein per kilogram of body weight (1 kg = 2.2 lb).

Step 8. Compute the percentage of your total calories supplied from carbohydrate, lipid, and protein.

> For example, if total average daily caloric intake is 2450 kcal/day, and 1600 kcal are from carbohydrates, the daily percentage of total calories from carbohydrates is: 1600/2450 × 100 = 65%

Step 9. While there is no specific RDA for lipid or carbohydrate, a prudent recommendation is that lipid should not exceed more than 30% of your total caloric intake; for active men and women, carbohydrates should be approximately 60% of the total calories ingested. Thus, in computing the percentage of "RDA" for graphing purposes in Figures E.1, you should assume that:
   a. "RDA" for lipid is 30% of total calories
   b. "RDA" for carbohydrate is 60% of total calories

> For example, if 50% of your average daily calories comes from lipid, you are taking in 167% of the recommended value ("RDA") for this nutrient: [50% divided by 30% (recommended percentage) × 100 = 167%]

Step 10. As was the case for lipid and carbohydrate, there is no RDA for average daily caloric intake. Any recommendation for energy intake needs to be based on one's present status for body fat as well as current daily energy expenditure. However, average values for daily caloric intake have been published for the typical young adult and equal about 2100 kcal for young women and 3000 kcal for young men. Thus, for graphing purposes in Figure E.1, you can evaluate your average daily caloric intake against the "average" values for your sex and age.

For example, if you are a 20-year-old female and you consume an average of 2400 kcal daily, your energy intake would be equal to 114% of the average ("RDA") for your age and sex. [2400 kcal divided by 2100 kcal (average) $\times$ 100 = 114%]. This does not mean that you need to go on a diet and reduce food intake to bring you in line with the average U.S. value. To the contrary, your higher-than-average caloric intake may be required to power your active lifestyle that contributes to maintaining a desirable body mass and body composition.

If you eat a food item not listed in Section IV, try to make an intelligent guess as to its composition and amount consumed. It is better to overestimate the amount of food consumed than to underestimate or to make no estimation at all. If you go to a restaurant for dinner, or to a friend's house where it may be inappropriate to measure the food, then omit this day from the counting procedure and resume record keeping the following day.

Because the purpose of keeping records for three days is to obtain an accurate appraisal of the average daily en-ergy and nutrient intake, record-keeping is extremely important. **Be sure to record everything you eat**. Most people find it easier to keep accurate records if they record food items while preparing a meal or immediately afterwards when eating snack items.

TABLE E.1. SAMPLE ONE-DAY CALORIC AND NUTRIENT INTAKE FOR A 21-YEAR-OLD COLLEGE STUDENT

Food Item	Amount	kcal	Protein (g)	CHO (g)	Lipid (g)	Ca (mg)	Fe (mg)	Fiber (g)	Cholesterol (mg)	Thiam[a] (mg)	Ribofl[a] (mg)
Breakfast											
Eggs, hard boiled	2 (2 oz ea)	160	14.1	1.4	11.2	55.2	1.9	0.0	452	0.06	0.53
Orange juice	8 oz	104	0.9	86.4	0.5	72.4	0.8	0.9	0.0	0.20	0.06
Corn flakes	1 cup/1 oz	110	2.3	24.4	0.5	1.0	1.8	0.6	0.0	0.37	0.42
Skim milk	8 oz	80	7.8	10.6	0.6	279.2	0.1	0.0	3.7	0.08	0.32
Snack											
None											
Lunch											
Tuna fish (oil pack)	2 oz	112	16.5	0.0	68.0	7.8	0.8	0.0	10.0	0.02	0.06
White bread (toast)	2 pieces	168	5.3	31.4	2.5	81.2	1.8	1.3	0.0	0.24	0.21
Mayonnaise	1 oz	203	0.3	0.8	22.6	5.7	0.2	0.0	16.8	0.01	0.01
Skim milk	8 oz	80	7.8	10.6	0.6	279.2	0.1	0.0	3.7	0.08	0.32
Plums	4 (2 oz ea)	128	1.8	29.5	1.4	10.3	0.2	4.4	0.0	0.10	0.22
Snack											
Chocolate milkshake	8 oz	288	7.7	46.4	8.4	256	0.7	0.3	29.6	0.13	0.55
Dinner											
Sirloin steak, lean	8 oz	456	64.8	0.0	20.2	18.6	5.8	0.0	173.6	0.21	0.47
French fries, veg. oil	6 oz	540	6.8	67.2	28.1	34.2	1.3	3.4	0.0	0.30	0.05
Cole slaw	4 oz	80	1.4	14.1	3.0	51.2	0.7	2.3	9.2	0.08	0.07
Italian bread	2 oz	156	5.1	32.0	1.0	9.4	1.5	0.9	0.0	0.23	0.13
Light beer	8 oz	96	0.6	8.8	0.0	11.2	0.1	0.5	0.0	0.02	0.06
Snack											
Yogurt, whole milk	6 oz	102	5.9	7.9	5.5	205.8	0.1	0.0	22.1	0.05	0.24
Daily Total		2863	149.1	371.5	174.1	1378.4	17.2	14.6	720.7	2.18	3.72

[a]Thiam, thiamin; Ribofl, riboflavin.

TABLE E.2.   DAILY AND AVERAGE DAILY SUMMARY CHART OF THE INTAKE OF CALORIES AND SPECIFIC FOOD NUTRIENTS

Day	kcal	Protein[a] (g)	Lipid[a] (g)	Carbo[a] (g)	Ca (mg)	Fe (mg)	Thiamine (mg)	Riboflavin (mg)	Fiber (g)	Cholesterol (mg)
#1										
#2										
#3										
Three-day total										
Average Daily Value[b]										

[a]Use the following caloric transformations to convert your average daily grams of carbohydrate (CHO), lipid, and protein to average daily calories:

    1 g CHO = 4 kcal
    1 g Lipid = 9 kcal
    1 g Protein = 4 kcal

[b]The Average Daily Value is used to determine the percentage of the RDA for your graph. See Table E.1 for sample calculations. Figure E.1 is a bar graph showing the nutrient values as a percentage of the average or recommended value for each item.

TABLE E.3.   RDA VALUES FOR SELECTED NUTRIENTS INCLUDING SAMPLE COMPUTATIONS FOR DERIVING THE PERCENT OF RDA FROM YOUR DIETARY SURVEY. VALUES LISTED IN THE TABLE ARE 100% VALUES FOR PURPOSES OF GRAPHING YOUR DIETARY DATA

MEN								
Age	kcal[a]	Protein (g/kg)	Ca (mg)	Fe (mg)	Thiamine (mg)	Riboflavin (mg)	Fiber[a] (g)	Cholesterol[a] (mg)
19–22	3000	0.8	1200	10	1.5	1.7	30	300
23–50	2700	0.8	800	10	1.5	1.7	30	300
WOMEN								
Age	kcal[a]	Protein (g/kg)	Ca (mg)	Fe (mg)	Thiamine (mg)	Riboflavin (mg)	Fiber[a] (g)	Cholesterol[a] (mg)
19–22	2100	0.8	1200	15	1.1	1.3	30	300
23–50	2000	0.8	800	15	1.1	1.3	30	300

Source: Recommended Dietary Allowances, Revised 1989. Washington, DC: Food and Nutrition Board, National Academy of Sciences– National Research Council, 1989.

[a]There is no RDA for daily caloric intake or for the intake of fiber or cholesterol. Values for caloric intake represent an average for adult Americans, while fiber and cholesterol values are recommended as being prudent for maintaining good health.

HOW TO DETERMINE THE PERCENTAGE OF THE RDA FROM YOUR DIETARY SURVEY

Example #1: Percentage of RDA for protein for a 70-kg person
              Daily protein intake = 68 g
              RDA = (70 kg × 0.8 g/kg) = 56 g
              % of RDA = 56/68 × 100 = 121%

Example #2: Percentage of RDA for iron (female)
              Daily iron intake = 7.5 mg
              RDA = 15 mg
              % of RDA = 7.5/15 × 100 = 50%

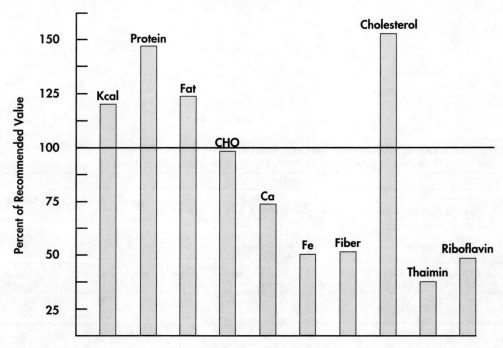

Figure E.1. Example of a bar graph to illustrate the food and nutrient intake expressed as a percentage of recommended values.

100% value represents:

kcal:	3000 kcal for men age 19–22
	2700 kcal for men age 23–50
	2100 kcal for women age 19–22
	2000 kcal for women age 23–50
Lipid:	30% of total calories
CHO:	60% of total calories
Fiber:	30 g
Cholesterol:	300 mg

Assessment of Physical Activity:
Three-Day Activity Recall

Any hope of changing the pattern and quantity of daily physical activity must be predicated on an accurate appraisal of the daily energy expenditure.

Step 1. Determine your daily pattern of physical activity, including such minimal daily requirements as sleeping, eating, sitting in class, and bathing. An activity profile can be constructed by keeping a daily log for three days of the actual time allotted to the various activities that represent your usual pattern of activity. To illustrate the procedure, Table F.1 shows a fairly detailed activity profile for a college professor during a typical day. This record includes a description of the activity, its duration, and the calories expended during the activity.

Step 2. (a) Determine your Basal Metabolic Rate (BMR) in kcal/h as follows:

> **MEN:**
> BMR (kcal/h) = 38 kcal/m^2/h^a × surface area (m^2)^b
>
> **WOMEN:**
> BMR (kcal/h) = 35 kcal/m^2/h^a × surface area (m^2)^b

^{a} You can also determine BMR (kcal/m^2/h) for your age and sex from Figure 7.2 in your textbook.
^{b} Surface area (m^2) is determined from stature and body mass; see How to, page 157 in your textbook.

(b) Divide the BMR value in kcal/h obtained in Step 2(a) by 60 to compute the BMR value per minute. This value will be used to represent your energy expenditure per minute during sleep.

> **EXAMPLE OF BMR CALCULATIONS**
> **DATA: MALE**
>
> Age, 40 y
> Stature, 182 cm (72 in)
> Body mass, 86.4 kg (190 lb)
> Body surface area (see How to, page 157) = 2.09 m^2
> kcal/m^2/h (from Fig. 7.2) = 37.2
>
> **CALCULATIONS**
>
> a. kcal/h = 37.2 × 2.09 = 77.7
> b. kcal/min = 77.7/60 = 1.3

Step 3. Determine the energy expenditure in kcal/min for each of the activities listed in your profile. For sleep use the BMR computed in Step 2b. These values are listed in Section IV of this Study Guide. The values are gross values that include the resting energy value. If an activity is not included, choose an activity most similar to the one you list.

Step 4. Multiply the energy expenditure value from the table by the number of minutes spent in each activity.

Step 5. Sum the total energy expenditure for each activity, including the value for sleep, to arrive at your total energy expenditure for the day.

Step 6. Repeat steps 2 through 5 for each of the three days. Obtain the average daily energy expenditure for the three-day pe-riod by summing the total calories expended over the three-day pe-riod and divide by 3.

HOW TO INTERPRET YOUR AVERAGE DAILY ENERGY EXPENDITURE

There is no norm or desirable standard for the number of calories you should expend during a day. Many factors are involved, including body size, age, sex, and—most importantly—level of physical activity. The average daily energy expenditure approximates 3000 kcal for men and 2100 kcal for women between the ages of 19 and 22 years. If you are not gaining or losing body weight, then your energy expenditure equals your energy intake.

TABLE F.1.   DETAILED RECORD OF PHYSICAL ACTIVITY FOR ONE DAY FOR A UNIVERSITY PROFESSOR

Activity	Begin Time	End Time	Total Minutes	Similar Activity Section IV	kcal/min	Total kcal
Wake, bathroom use	6:45AM	6:53AM	8	Standing quietly	2.3	18.4
Go back to bed	6:53	7:30	37	BMR	1.3	48.1
Eat breakfast	7:30	7:50	10	Eating, sitting	2.0	40.0
Use bathroom	7:50	8:00	10	Standing quietly	2.3	23.0
Dress	8:00	8:06	6	Standing quietly	2.3	13.8
Drive to school	8:06	8:17	11	Sitting quietly	2.0	22.0
Walk to office	8:17	8:25	8	Walking, normal pace	6.9	55.8
Work in office, pick up mail	8:25	10:00	95	Writing, sitting	2.5	237.5
Up/down stairs	10:00	10:10	10	11 min 30 sec pace	11.7	117.0
Work in office	10:10	12:10PM	120	Writing, sitting	2.5	300.0
Go to locker	12:10PM	12:12	2	Walking, normal pace	6.9	13.8
Get dressed	12:12	12:16	4	Standing quietly	2.3	9.2
Walk to track	12:16	12:20	4	Walk, normal pace	6.9	27.6
Wait for friend	12:20	12:30	10	Standing quietly	2.3	23.0
Run to park, back	12:30	2:00	90	8-min mile pace	17.2	1553.0
Walk to locker	2:00	2:04	4	Walk, normal pace	6.9	27.6
Shower, dress	2:04	2:20	16	Quiet standing	2.3	36.8
Walk to office	2:20	2:24	4	Walk, normal pace	6.9	27.6
Meeting/lunch	2:24	3:00	36	Eating, sitting	2.0	72.0
Work in office	3:00	5:05	125	Writing, sitting	2.5	312.5
Walk to library	5:05	5:12	7	Walk, normal pace	6.9	48.3
Work in library	5:12	6:05	53	Writing, sitting	2.5	132.5
Walk to dean	6:05	6:10	5	Walk, normal pace	6.9	34.5
Meeting, dean	6:10	6:35	25	Writing, sitting	2.5	62.5
Walk to office	6:35	6:43	8	Walk, normal pace	6.9	55.2
Walk to car	6:43	6:51	8	Walk, normal pace	6.9	55.2
Drive home	6:51	7:03	12	Sitting quietly	1.8	21.6
Change clothes	7:03	7:07	4	Standing quietly	2.3	9.2
Wash-up	7:07	7:11	4	Standing quietly	2.3	9.2
Cook dinner	7:11	8:00	49	Cooking	4.1	200.9
Watch TV	8:00	8:30	30	Sitting quietly	1.8	54.0
Eat dinner	8:30	9:00	30	Eating, sitting	2.0	60.0
Mail letter	9:00	9:05	5	Walk, normal pace	6.9	34.5
Listen to stereo	9:05	9:30	25	Sitting quietly	1.8	45.0
Watch TV	9:30	10:30	60	Sitting quietly	1.8	108.0
Wash-up	10:30	10:38	8	Standing quietly	2.3	18.4
Read in bed	10:38	11:15	37	Lying at ease	1.9	70.3
Sleeping	11:15	6:45	460	BMR	1.3	598

Assessment of Health-Related Physical Fitness

The trend in fitness assessment over the past 20 years has been to deemphasize physical fitness tests that stress motor performance and athletic fitness (i.e., tests of speed, power, balance, and agility). Today's trend is to focus on those measures that assess functional capacity and also reflect various aspects of overall good health or disease prevention, or both. The four components of health-related physical fitness that are commonly evaluated include aerobic power (cardiorespiratory fitness), body composition, abdominal muscular strength and endurance, and lower back and hamstring flexibility. An upper-extremity measure of muscular strength also is often included, although it is not directly related to health status.

A person's performance on each of the test items of health-related physical fitness is not fixed but can be significantly improved through a program of regular exercise and weight control.

TEST COMPONENT #1: LOWER-BACK FLEXIBILITY

A. Rationale

A substantial amount of clinical evidence indicates that maintenance of trunk, hip, lower back, and posterior thigh flexibility is important in the prevention and alleviation of lower-back pain and tension. Back disorders are associated with weak muscles in the abdominal area and limited range of motion of the lower spine. There also is reduced elasticity of the hamstring muscles at the back of the upper leg. The importance of lower-back flexibility as a health-related fitness measure is further supported by the fact that physicians frequently prescribe specific trunk and thigh flexibility stretches for their patients with lower-back problems, or for individuals who desire to prevent the occurrence of such problems.

B. Lower-Back Flexibility Assessment Test: The Sit and Reach Test

Sit on a mat with your legs extended. Your feet should rest against the base of a box on which a yardstick is mounted with the 9 inch (23 cm) mark on the near side of the box (see Fig. G.1). After a general warm-up that includes stretching of the lower back and posterior thighs, slowly reach forward with both hands as far as possible and hold the position momentarily. Record the distance reached on the yardstick by your fingertips. Use the best of four trials as your flexibility score.

C. How Do You Rate on Lower-Back Flexibility?

Compare your results with the normative data in Table G.1.

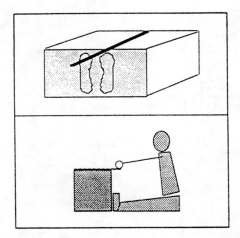

Figure G.1.   Sit and reach flexibility test.

TABLE G.1.   NORMS FOR THE SIT AND REACH (CM) FOR YOUNG ADULT MEN AND WOMEN

Percentile	Men				Women			
	18 y	19 y	20 y	21 y	18 y	19 y	20 y	21 y
99	50	49	49	50	52	52	51	50
95	45	45	46	45	47	47	46	46
90	42	43	43	42	46	45	45	44
85	41	42	41	41	44	43	43	43
80	40	40	41	40	43	42	42	42
75	39	39	40	39	42	41	41	42
70	38	38	39	38	40	41	39	40
65	37	37	38	36	40	40	38	39
60	36	36	37	35	39	38	38	38
55	35	35	36	35	38	38	37	37
50	34	34	35	33	38	37	37	36
45	34	33	34	32	37	36	36	36
40	32	32	33	31	36	36	35	35
35	31	31	32	31	35	34	34	34
30	30	29	31	30	34	33	33	33
25	29	28	30	28	33	32	32	32
20	27	27	27	27	32	31	31	31
15	25	26	25	25	30	29	30	29
10	23	23	22	24	29	27	28	27
5	19	19	18	20	26	23	24	25

From: Norms for College Students: Health Related Physical Fitness Test. Reston, Va, American Alliance for Health, Physical Education, Recreation, and Dance, 1985.

TEST COMPONENT #2: BODY FAT PERCENTAGE

A. Rationale

Body composition is defined as the relative percentage of fat and fat-free body mass. An excessive accumulation of body fat significantly hinders the ability to perform tasks requiring speed, endurance, and coordination. Additionally, research indicates that an excess amount of body fat is an associative and/or contributing factor to four categories of health hazards: (*a*) disturbance of normal body functions, (*b*) increased risk of disease (e.g., hypertension, high cholesterol, diabetes, coronary heart disease), (*c*) exacerbation of existing disease states, and (*d*) adverse psychological effects.

B. Body Fat Percentage Assessment Test: Measurement of Fatfolds

Evaluation of body fat can be made by different indirect procedures. These can include *girths* or *fatfolds* measured at specific body sites. In the *Essentials of Exercise Physiology* textbook we describe how to use girth measurements to estimate percentage body fat. In this section we describe an alternative method using skinfolds. The skinfold measurements are:

> **MEN:** Chest, Abdomen, Thigh
> **WOMEN:** Triceps, Suprailium, Thigh

Measurements at each site are taken in accordance with standard procedures. The exact anatomical sites for each skinfold measurement are described in chapter 18 of your textbook.

> **Abdomen** *(Men only):* Vertical fold measured 1 inch to the right of the umbilicus.
> **Chest** *(Men only):* A diagonal fold taken one-half of the distance between the anterior axillary line and the nipple.
> **Suprailium** *(Women only):* Slightly oblique fold just above the crest of the hip bone. The fold is lifted to follow the natural diagonal line at this point.
> **Thigh** *(Men and Women):* Vertical fold measured at the anterior midline of the thigh, midway between the knee cap and the hip.
> **Triceps** *(Women only):* Vertical fold measured at the posterior midline of the upper arm halfway between the tip of the shoulder and the tip of the elbow. The elbow should be extended and relaxed.

Compute your body fat percentage from the sum of the three fatfold measures as indicated in Table G.2 (men) or Table G.3 (women). Then see how you rate by determining your percent body fat percentile using the data in Table G.4.

TABLE G.2. PERCENT FAT ESTIMATES FOR MEN. SUM OF CHEST, ABDOMINAL, AND THIGH SKINFOLDS

Sum Skinfolds (mm)	AGE TO THE LAST YEAR			
	Under 22	23 to 27	28 to 32	33 to 37
8–10	1.3	1.8	2.3	2.9
11–13	2.2	2.8	3.3	3.9
14–16	3.2	3.8	4.3	4.8
17–19	4.2	4.7	5.3	5.8
20–22	5.1	5.7	6.2	6.8
23–25	6.1	6.6	7.2	7.7
26–28	7.0	7.6	8.1	8.7
29–31	8.0	8.5	9.1	9.6
32–34	8.9	9.4	10.0	10.5
35–37	9.8	10.4	10.9	11.5
38–40	10.7	11.3	11.8	12.4
41–43	11.6	12.2	12.7	13.3
44–46	12.0	13.1	13.6	14.2
47–49	13.4	13.9	14.5	15.1
50–52	14.3	14.8	15.4	15.9
53–55	15.1	15.7	16.2	16.8
56–58	16.0	16.5	17.1	17.7
59–61	16.9	17.4	17.9	18.5
62–64	17.6	18.2	18.8	19.4
65–67	18.5	19.0	19.6	20.2
68–70	19.3	19.9	20.4	21.0
71–73	20.1	20.7	21.2	21.8
74–76	20.9	21.5	22.0	22.6
77–79	21.7	22.2	22.8	23.4
80–82	22.4	23.0	23.6	24.2
83–85	23.2	23.8	24.4	25.0
86–88	24.0	24.5	25.1	25.7
89–91	24.7	25.3	25.9	25.5
92–94	25.4	26.0	26.6	27.2
95–97	26.1	26.7	27.3	27.9
98–100	26.9	27.4	28.0	28.6
101–103	27.5	28.1	28.7	29.3
104–106	28.2	28.8	29.4	30.0
107–109	28.9	29.5	30.1	30.7
110–112	29.6	30.2	30.8	31.4
113–115	30.2	30.8	31.4	32.0
116–118	30.9	31.5	32.1	32.7
119–121	31.5	32.1	32.7	33.3
122–124	32.1	32.7	33.3	33.9
125–127	32.7	33.3	33.9	34.9

From: Jackson, A. S. and Pollock, M. L. Practical Assessment of Body Composition. Phys. Sportsmed., 13 (5): 76, 1985. Reproduced with permission of McGraw-Hill, Inc.

TABLE G.3. PERCENT FAT ESTIMATES FOR WOMEN. SUM OF TRICEPS, SUPRAILIAC, AND THIGH SKINFOLDS

Sum Skinfolds (mm)	AGE TO THE LAST YEAR			
	Under 22	23 to 27	28 to 32	33 to 37
23–25	9.7	9.9	10.2	10.4
26–28	11.0	11.2	11.5	11.7
29–31	12.3	12.5	12.8	13.0
32–34	13.6	13.8	14.0	14.3
35–37	14.8	15.0	15.3	15.5
38–40	16.0	16.3	16.5	16.7
41–43	17.2	17.4	17.7	17.9
44–46	18.3	18.6	18.8	19.1
47–49	19.5	19.7	20.0	20.2
50–52	20.6	20.8	21.1	21.3
53–55	21.7	21.9	22.1	22.4
56–58	22.7	23.0	23.2	23.4
59–61	23.7	24.0	24.2	24.5
62–64	24.7	25.0	25.2	25.5
65–67	25.7	25.9	26.2	26.4
68–70	26.6	26.9	27.1	27.4
71–73	27.5	27.8	28.0	28.3
74–76	28.4	28.7	28.9	29.2
77–79	29.3	29.5	29.8	30.0
80–82	30.1	30.4	30.6	30.9
83–85	30.9	31.2	31.4	31.7
86–88	31.7	32.0	32.2	32.5
89–91	32.5	32.7	33.0	33.2
92–94	33.2	33.4	33.7	33.9
95–97	33.9	34.1	34.4	34.6
98–100	34.6	34.8	35.1	35.3
101–103	35.3	35.4	35.7	35.9
104–106	35.8	36.1	36.3	36.6
107–109	36.4	36.7	36.9	37.1
110–112	37.0	37.2	37.5	37.7
113–115	37.5	37.8	38.0	38.2
116–118	38.0	38.3	38.5	38.8
119–121	38.5	38.7	39.0	39.2
122–124	39.0	39.2	39.4	39.7
125–127	39.4	39.6	39.9	40.1
128–130	39.8	40.0	40.3	40.5

From: Jackson, A. S. and Pollock, M. L. Practical Assessment of Body Composition. Phys. Sportsmed. 13 (5): 76, 1985. Reproduced with permission of McGraw-Hill, Inc.

TABLE G.4. PERCENTILE NORMS FOR PERCENT BODY FAT FOR MEN AND WOMEN[a]

Percentile	Males	Females
95	6.7	16.8
90	8.1	18.1
85	9.8	19.8
80	10.8	20.8
75	11.6	21.6
70	12.4	22.4
65	13.1	23.1
60	13.7	23.7
55	14.4	24.4
50	15.0	25.0
45	15.6	25.6
40	16.3	26.3
35	16.9	26.9
30	17.6	27.6
25	18.4	28.4
20	19.2	29.2
15	20.2	30.2
10	21.9	31.9

From McArdle, W. D., et al.: Exercise Physiology: Energy, Nutrition, and Human Performance. 3rd Edition. Philadelphia: Lea & Febiger, 1991.
[a]Normative standards are based on an average body fat for males of 15% and 25% for females with one standard deviation of 6 5% body fat for both sexes.

TEST COMPONENT #3: AEROBIC/CARDIOVASCULAR FUNCTION

A. Rationale

Enhanced aerobic-cardiovascular function permits higher levels of extended energy expenditure and physical working capacity, and also facilitates recovery. Also, lack of regular exercise (with a reduced aerobic fitness) has been linked to an increased risk of heart disease.

B. Aerobic-Cardiovascular Assessment Test: The 1-Mile Run

Aerobic-cardiovascular performance during exercise can be measured by running a distance of 1 mile. Warm up for several minutes, then run/walk as rapidly as possible for 1 mile, and record your time to the nearest second. This test should be performed on a quarter-mile (440-yd) track.

C. How Do You Rate on Aerobic-Cardiovascular Function?

Compare your mile run times with the normative data in Table G.5.

TABLE G.5.   PERCENTILE NORMS FOR THE MILE RUN (MIN:SEC) FOR AGE AND SEX

Percentile	Men (age, y)				Women (age, y)			
	18	19	20	21	18	19	20	21
99	4:57	5:00	4:33	4:38	5:33	5:27	5:16	6:26
95	5:29	5:30	5:21	5:28	7:01	6:56	7:00	7:02
90	5:43	5:42	5:40	5:47	7:28	7:22	7:21	7:21
85	5:55	5:55	5:46	6:01	7:47	7:45	7:41	7:35
80	6:05	6:04	5:59	6:09	8:01	8:00	7:59	7:47
75	6:13	6:09	6:08	6:15	8:15	8:13	8:15	8:01
70	6:21	6:15	6:15	6:22	8:31	8:25	8:30	8:09
65	6:28	6:25	6:23	6:31	8:48	8:34	8:40	8:16
60	6:35	6:32	6:30	6:39	8:59	8:52	8:56	8:30
55	6:40	6:39	6:35	6:47	9:06	9:01	9:07	8:40
50	6:48	6:45	6:43	6:53	9:23	9:13	9:20	8:57
45	6:55	6:53	6:51	6:57	9:35	9:26	9:29	9:04
40	7:03	7:01	6:56	7:02	9:49	9:41	9:43	9:24
35	7:10	7:09	7:09	7:09	10:01	10:00	9:45	9:51
30	7:17	7:15	7:19	7:20	10:16	10:07	10:10	10:02
25	7:29	7:27	7:29	7:34	10:35	10:25	10:28	10:30
20	7:45	7:45	7:41	7:49	10:50	11:00	10:45	11:00
15	8:05	8:00	7:56	8:07	11:17	11:18	11:00	11:20
10	8:33	8:30	8:30	8:29	12:00	12:00	11:30	12:13
5	9:40	9:31	9:48	9:22	13:01	13:05	12:39	12:57

From: Norms for College Students: Health Related Physical Fitness Test. Reston, VA, American Alliance For Health, Physical Education, Recreation, and Dance, 1985.

TEST COMPONENT #4: ABDOMINAL MUSCULAR STRENGTH AND ENDURANCE

A. Rationale

Abdominal muscular strength and endurance are important in stabilizing the torso while performing diverse physical tasks. From a health-related perspective, the functional capacity of this muscle group is of considerable importance to the maintenance of proper posture and spinal alignment, and to provide muscular support to reduce lower back strain. Clinical evidence implicates weak muscles of the abdominal wall as a major cause of lower back pain; strengthening these muscles is consistently recommended in both preventive and rehabilitative back programs.

B. Abdominal Muscular Strength and Endurance Assessment: The Modified Sit-Up Test

Lie on your back with your knees flexed, feet flat on floor, and heels between 12 and 18 inches from the buttocks (see Fig. G.2). Cross your arms over your chest with the hands on opposite shoulders. Have a partner hold your feet to keep them in touch with the floor. Curl to the sitting position; arm contact with the chest must be maintained, and the chin should remain tucked to the chest. The sit-up is completed when your elbows touch your thighs. Return to the starting position until your mid-back contacts the floor.

Your partner gives the signal "Ready, Go." The test is started on the word "Go" and ceases on the word "Stop." Your score is the number of correctly executed sit-ups performed in 60 seconds. If necessary, you can rest during the test.

C. How Do You Rate on Abdominal Muscular Strength and Endurance?

Compare your abdominal muscular strength and endurance with the normative data in Table G.6.

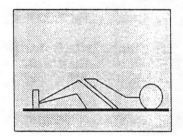 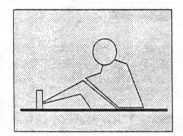

Figure G.2. Modified sit-up.

TABLE G.6. PERCENTILE NORMS FOR 1-MINUTE TIMED SIT-UPS FOR AGE AND SEX

Percentile	Men (age, y)				Women (age, y)			
	18	19	20	21	18	19	20	21
99	70	69	65	67	61	62	63	67
95	62	60	54	61	54	52	53	54
90	57	57	55	55	49	48	49	51
85	55	55	52	54	46	45	46	47
80	53	52	52	52	44	43	44	46
75	51	50	50	50	42	41	42	45
70	50	49	49	48	40	40	41	42
65	48	48	48	47	39	39	40	40
60	47	46	47	45	38	38	38	39
55	46	45	46	44	37	37	37	37
50	45	44	45	43	35	36	36	36
45	43	43	44	42	34	35	35	35
40	42	42	43	41	33	33	34	34
35	41	41	41	41	32	32	33	33
30	40	40	40	40	31	31	31	31
25	39	39	39	39	30	30	30	30
20	37	36	37	37	29	29	30	29
15	36	35	35	36	27	27	28	26
10	34	33	33	34	25	25	25	25
5	30	30	30	31	21	22	22	21

From: Norms For College Students: Health Related Physical Fitness Test. Reston, VA, American Alliance For Health, Physical Education, Recreation, and Dance, 1985.

TEST COMPONENT #5: UPPER-BODY MUSCULAR STRENGTH

A. Rationale

While not a bona-fide health-related fitness component, upper-body strength (shoulders, chest, upper arms) is required for the performance of many sport, recreational, and occupational tasks. The 1-RM bench press measures the maximum weight pushed from the bench press position to full extension of the arms, and is used as a measure of upper-body strength. The test is scored as a ratio of weight pushed divided by your body weight.

B. Upper-Body Muscular Strength Assessment: The 1-RM Bench Press Ratio Test

Equipment includes either a barbell set with a bench, or a fixed bench press station on a single or multi-station apparatus. Determining the maximum weight lifted in one repetition by trial and error. Find a weight you can press two or three times. Add increments of 10 pounds and attempt the lift again. Add increments of 5 or 10 pounds (with rest between lifts) until you determine the maximum amount you can lift one time.

C. How Do You Rate on Upper-Body Muscular Strength?

Compare your results with the normative standards in Table G.7 that have been developed in our university strength and conditioning classes.

TABLE G.7. UPPER-BODY BENCH PRESS TO BODY WEIGHT RATIO NORMS (WEIGHT PUSHED [LB] DIVIDED BY BODY WEIGHT [LB])

	Female (age, y)		Male (age, y)	
Percentile	18–20	20–29	18–20	20–29
99	>0.90	>1.06	>1.70	>1.60
95	0.86	1.00	1.58	1.55
90	0.83	0.92	1.42	1.44
85	0.80	0.88	1.38	1.30
80	0.75	0.83	1.32	1.29
75	0.72	0.79	1.26	1.26
70	0.68	0.76	1.24	1.20
65	0.70	0.74	1.20	1.14
60	0.67	0.70	1.16	1.11
55	0.64	0.68	1.14	1.07
50	0.63	0.64	1.11	1.06
45	0.61	0.61	1.08	1.00
40	0.60	0.59	1.06	0.97
35	0.58	0.57	1.00	0.96
30	0.55	0.56	0.98	0.90
25	0.54	0.54	0.90	0.85
20	0.53	0.51	0.86	0.82
15	0.52	0.47	0.82	0.80
10	0.47	0.46	0.81	0.72
5	0.41	0.42	0.76	0.70
1	<0.39	<0.40	<0.68	<0.65

YOUR OVERALL HEALTH-RELATED FITNESS PROFILE

Create a bar graph to visually display your percentile ranking on each component of your health-related fitness. Because scores on each item are not strongly related to each other (i.e., people with good muscular strength do not necessarily score high in aerobic fitness or flexibility, etc.) it is unusual for individuals to rank high on all items.

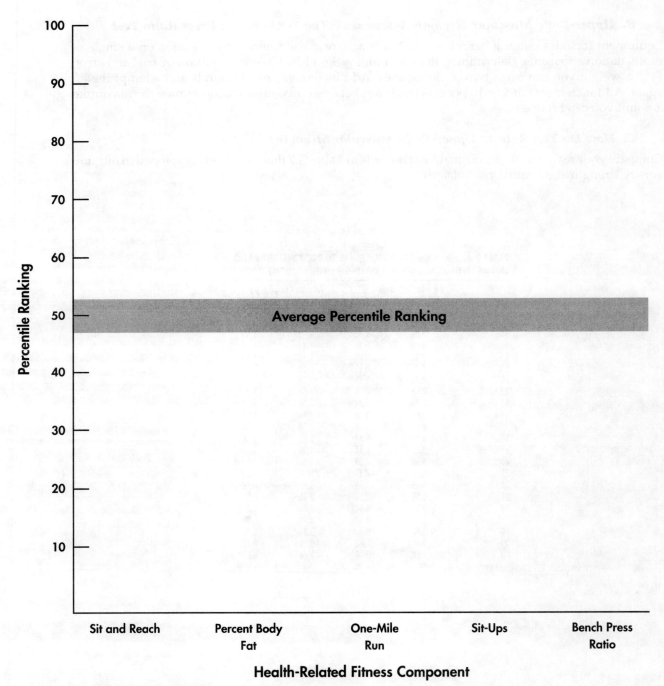

Figure G.3. Health-related fitness profile.

Assessment of Heart Disease Risk (RISKO)

The assessment of heart disease risk factors provides some idea of your chances for developing heart disease. The chart on the next page is modified from a more elaborate version generated from 35 years of research on the natural history of heart disease in Framingham, Massachusetts. RISKO is appropriate for adult men and women of all ages. While a RISKO score is certainly not a substitute for a regular medical checkup, the information is helpful in providing insight as to potential areas for concern. Many of these telltale characteristics are habits (or the result of habits) that can be controlled. It certainly would be beneficial to identify risk factors at an early age to thwart the escalation of silent heart disease so prevalent in our highly mechanized, sedentary society.

Assign the appropriate numerical value that represents your present status for each category. Find the box applicable to you and circle the number in it. For example, if you are 19 years old, circle the number "1 pt" in the box labeled 10 to 20 years. After checking all the rows, add the circled numbers. The total number of points is your risk score. Refer to the **Relative Risk Category** to see how you rank. While there is nothing you can do about your age, sex, and heredity, other risks such as high blood pressure, tension, cigarette smoking, serum cholesterol, diet, lack of exercise, and obesity can be modified if not totally eliminated!

Explanation of Risk Variables

Heredity

Count parents, brothers, and sisters who have had a heart attack or stroke.

Smoking

If you inhale deeply and smoke a cigarette way down, add one point to your score. Do not subtract because you think you do not inhale or smoke only a half inch on a cigarette.

Exercise

Lower your score one point if you exercise regularly and frequently.

Cholesterol–Saturated Fat Intake

A cholesterol blood level is best. If you have not had a blood test recently, then estimate honestly the percentage of solid fats you eat. These are usually of animal origin like lard, cream, butter, and beef and lamb fat. If you eat much saturated fat, your cholesterol level will probably be high.

Blood Pressure

If you have no recent reading but have passed an insurance or general medical examination, chances are you have a systolic blood pressure level (upper reading) of 140 or less.

Sex

This takes into account the fact that men have from 6 to 10 times more heart attacks than women of childbearing age.

HEREDITY	No known history of heart disease **1 pt**	1 relative with cardiovascular disease over age 60 **2 pts**	2 relatives with cardiovascular disease over age 60 **3 pts**	1 relative with cardiovascular disease under age 60 **4 pts**	2 relatives with cardiovascular disease under age 60 **6 pts**	3 relatives with cardiovascular disease under age 60 **8 pts**
AGE	10 to 20 y **1 pt**	21 to 30 y **2 pts**	31 to 40 y **3 pts**	41 to 50 y **4 pts**	51 to 60 y **6 pts**	61 and over **8 pts**
CHOLESTEROL OR DIETARY FAT %	Cholesterol below 180 mg/dL; diet contains no animal or solid fats **1 pt**	Cholesterol 180–205 mg/dL; diet contains 10% animal or solid fats **2 pts**	Cholesterol 206–230 mg/dL; diet contains 20% animal or solid fats **3 pts**	Cholesterol 231–255 mg/dL; diet contains 30% animal or solid fats **4 pts**	Cholesterol 256–280 mg/dL; diet contains 40% animal or solid fats **5 pts**	Cholesterol 281–300 mg/dL; diet contains 50% animal or solid fats **7 pts**
SEX	Female under age 40 y **1 pt**	Female age 40 to 50 y **2 pts**	Female over age 50 y **3 pts**	Male **4 pts**	Stocky male **6 pts**	Bald stocky male **7 pts**
EXERCISE	Intensive occupational and recreational exertion **1 pt**	Moderate occupational and recreational exertion **2 pts**	Sedentary work and intense recreational exertion **3 pts**	Sedentary occupational and moderate recreational exertion **5 pts**	Sedentary work and light recreational exertion **6 pts**	Complete lack of all exercise **3 pts**
BLOOD PRESSURE	100 upper reading **1 pt**	120 upper reading **2 pts**	140 upper reading **3 pts**	160 upper reading **4 pts**	180 upper reading **6 pts**	200 or more upper reading **8 pts**
TOBACCO SMOKING	Non-user **0 pts**	Cigar and/or pipe **1 pts**	10 cigarettes or less per day **3 pts**	20 cigarettes per day **4 pts**	30 cigarettes per day **6 pts**	40 cigarettes per day **8 pts**
BODY WEIGHT	+5 lb below standard weight **0 pts**	−5 to +5 lb of standard weight **1 pt**	6 to 20 lb overweight **2 pts**	21 to 35 lb overweight **3 pts**	36 to 50 lb overweight **5 pts**	51 to 65 lb overweight **7 pts**

YOUR SCORE _____ pts

SCORE	**RELATIVE RISK CATEGORY**	**SCORE**	**RELATIVE RISK CATEGORY**
6 to 11	Risk well below average	**25 to 31**	Moderate risk
12 to 17	Risk below average	**32 to 40**	High risk
18 to 24	Average risk	**41 to 62**	Very high risk, see a doctor

Desirable Body Weight and Body Fat Distribution

There are a number of different methods used to determine desirable body weight. We present four of the most common methods. **Method One** is based on the popular Height-Weight Charts. **Method Two** is based on your Body Mass Index, which is computed as the ratio of your body weight to height squared. **Method Three** determines your desirable body weight based on your percent body fat. **Method Four** does not determine your desirable body weight, but rather provides an estimate of your desirable body fat distribution based on the ratio of your waist girth to your hip girth.

METHOD ONE: HEIGHT–WEIGHT TABLES

1. Determine your height to the nearest half-inch.
2. Determine your body weight to the nearest one half-pound.
3. Review Table I.1 that presents weight-for-height for males and females.
4. Determine your desirable weight range as the midpoint and the low-end value of the listed weight range. The upper value of your desirable weight range would be your current body weight minus the midpoint of the desirable weight range from Table I.1. The low end of your desirable weight range would be your current body weight minus the low end of your desirable weight range from Table I.1.

CALCULATIONS

_____ Height (in)

_____ Body Weight (lb)

_____ Desirable Body Weight Range (From Table I.1)

_____ **Minus** _____ **Equals** _____
Current Body Weight | Midpoint of Desirable Range | Upper Value of Desirable Range

_____ **Minus** _____ **Equals** _____
Current Body Weight | Low End of Desirable Range | Lower Value of Desirable Range

TABLE I.1. USE WITH METHOD ONE. WEIGHT RANGES FOR HEIGHT FOR MALES AND FEMALES

Height (without shoes)		Weight—Males		Weight—Females	
cm	ft-in	kg	lb	kg	lb
146	4'9"			48–54	106–118
149	4'10"			48–54	106–120
151	4'11"			50–56	110–123
154	5'0"			51–57	112–126
156	5'1"	57–62	126–136	52–59	115–129
159	5'2"	58–63	128–138	54–60	118–132
162	5'3"	59–64	130–140	55–61	121–135
164	5'4"	60–65	132–143	56–63	124–138
167	5'5"	61–66	134–146	58–64	127–141
169	5'6"	62–68	137–149	59–65	130–144
172	5'7"	64–69	140–152	60–67	133–147
174	5'8"	65–70	143–155	62–68	136–150
177	5'9"	66–72	146–158	63–69	139–153
179	5'10"	68–73	149–161	64–71	142–156
182	5'11"	69–75	152–165		
185	6'0"	70–77	166–169		
187	6'1"	72–78	168–173		
190	6'2"	73–80	162–177		
192	6'3"	76–83	166–182		

METHOD TWO: BODY MASS INDEX (BMI)

1. Determine your height in meters. (Multiply your height in inches by 0.0254 to determine your height in meters.)
2. Determine your body weight in kilograms. (Divide your body weight in pounds by 2.205 to determine your weight in kilograms.)
3. Determine your BMI as weight (in kilograms) divided by height (in meters) squared.
4. Determine the desirable body weight corresponding to the upper and lower limits of the desirable BMI values as:

$$\text{Desirable Body Weight (kg)} = \text{BMI} \times [\text{Height (m)}]^2$$

The suggested BMI ranges for females is 21.3 to 22.1, and for males it is 21.9 to 22.4. BMI values above 27.8 for men and 27.3 for women have been associated with increased incidence rates of high blood pressure, diabetes, and coronary heart disease. The American Dietetic Association, in its position statement on nutrition and physical fitness, states that a BMI of 30 or more is classified as obese. (See Chapter 19 in your textbook for more details.)

CALCULATIONS

_____ Height, m (in $\times$ 0.0254 = m)

_____ Body Weight, kg (lb/2.205 = kg)

_____ BMI = Body Weight, kg/(Height, m)^2

_____ Body Weight at Upper End of Desirable BMI Range
Males: Body Weight, kg = 22.4 $\times$ (Height, m)^2
Females: Body Weight, kg = 22.1 $\times$ (Height, m)^2

_____ Body Weight at Lower End of Desirable BMI Range
Males: Body Weight, kg = 21.9 $\times$ (Height, m)^2
Females: Body Weight, kg = 21.3 $\times$ (Height, m)^2

EXAMPLE

Male: **Height** = 5'9" (1.753 m); **Weight** = 170 lb (77.1 kg); **BMI** = 25.089 kg/m^2 (77.1/1.753^2)
Desirable Body Weight at Upper End of Desirable BMI Range = (22.4 $\times$ 1.753^2) = 68.84 kg (151.8 lb)
Desirable Body Weight at Lower End of Desirable BMI Range = (21.9 $\times$ 1.753$-$) = 67.30 kg (148.4 lb)

   In this example, this person should lose 19 to 22 lb to bring his body weight into the desirable upper range of BMI values.

METHOD THREE: FAT-FREE BODY MASS (FFM)

For this method you will need to determine your percentage body fat. You can do this by using either the skinfold, girth, or underwater weighing method. Refer to Chapter 18 in your textbook on how to do these calculations. Also, self-assessment test G in this Study Guide presents the methods for predicting your percentage body fat by the skinfold technique.

STEP 1. Determine your body fat percentage from skinfold, girths, or other method
STEP 2. Determine your body's fat weight as body weight times body fat percentage expressed as a decimal (25% body fat would be 0.25)
STEP 3. Determine your fat-free body mass as body mass minus fat weight
STEP 4. Indicate the body percentage fat that you desire
STEP 5. Determine your desired body weight as:

$$\text{Desired Body Weight} = \frac{\text{FFM}}{1.00 - \text{Desired \% Body Fat}}$$
$$\text{(expressed as a decimal)}$$

CALCULATIONS

_____ Percentage body fat from skinfolds, girths, or other method

_____ Body fat weight, kg = body weight $\times$ decimal %body fat

_____ FFM, kg = body weight minus fat weight

_____ Desired body weight at desired %body fat = (FFM/1.00 − desired %body fat)

EXAMPLE

Body weight	= 190 lb (86.17 kg)
Percent body fat	= 19%
Fat weight	= 36.1 lb (16.37 kg)
FFM	= 153.9 lb (69.80 kg)
Desired body weight	= 153.9 lb/(1.00 − 0.15)
	= 153.9 lb/0.85
	= 181.1 lb (82.1 kg)
Recommended fat	= 190 lb − 181.1 lb fat loss
Loss	= 8.9 lb (4.04 kg)

METHOD FOUR: WAIST-TO-HIP RATIO (WHR)

The waist-to-hip girth ratio (WHR) is a measure of regional fat distribution. It is obtained by measuring the waist girth (average of the girth at the natural waist and at the umbilicus) and the hip girth (largest girth around the hips or buttocks). Wear tight clothing or no clothing. Do not compress the skin while taking the measurements. The significance of the WHR as a health risk is discussed in Chapter 19 of your textbook.

STEP 1. Determine your waist girth in centimeters

STEP 2. Determine your hip girth in centimeters

STEP 3. Determine your waist-to-hip girth ratio (waist girth/hip girth) (round to 2 significant digits; e.g., 0.90)

STEP 4. Compare your results with the recommendations below

CALCULATIONS

_____ Waist girth (average of girth at waist and girth at navel)

_____ Hip girth (widest girth around hips)

_____ WHR (average waist girth/hip girth)

EXAMPLE

FEMALE

1. Natural waist girth = 90.6 cm
2. Umbilicus girth = 64.5 cm
3. Average waist girth = (90.6 + 64.5/2) = 77.55 cm
4. Hip girth = 86.5 cm
5. WHR = 77.5 cm/86.5 cm = 0.90
6. Risk = Higher risk

WAIST-TO-HIP GIRTH RATIO RISK RATINGS

	Males	Females
Higher risk	>0.90	>0.85
Moderately high risk	0.90–0.95	0.80–0.85
Lower risk	<0.90	<0.80

Section III

- Answers to Chapter Quizzes
- Crossword Puzzle Solutions

Answers to Chapter Quizzes

CHAPTER 1

Multiple choice		True/False	
1.	a	1.	T
2.	c	2.	F
3.	c	3.	T
4.	a	4.	T
5.	b	5.	T
6.	b	6.	F
7.	b	7.	T
8.	a	8.	T
9.	a	9.	F
10.	b	10.	F

CHAPTER 2

Multiple choice		True/False	
1.	c	1.	T
2.	a	2.	T
3.	c	3.	T
4.	c	4.	F
5.	a	5.	T
6.	c	6.	T
7.	a	7.	F
8.	b	8.	T
9.	c	9.	T
10.	b	10.	F

CHAPTER 3

Multiple choice		True/False	
1.	b	1.	T
2.	a	2.	F
3.	b	3.	T
4.	c	4.	T
5.	a	5.	T
6.	b	6.	F
7.	a	7.	T
8.	a	8.	F
9.	a	9.	F
10.	a	10.	T

CHAPTER 4

Multiple choice		True/False	
1.	b	1.	F
2.	a	2.	T
3.	b	3.	T
4.	a	4.	F
5.	d	5.	F
6.	c	6.	F
7.	e	7.	F
8.	c	8.	T
9.	a	9.	T
10.	a	10.	T

CHAPTER 5

Multiple choice		True/False	
1.	b	1.	T
2.	a	2.	F
3.	a	3.	F
4.	d	4.	F
5.	e	5.	T
6.	b	6.	F
7.	d	7.	T
8.	b	8.	F
9.	a	9.	F
10.	d	10.	T

CHAPTER 6

Multiple choice		True/False	
1.	c	1.	T
2.	b	2.	F
3.	b	3.	F
4.	a	4.	T
5.	e	5.	T
6.	c	6.	F
7.	a	7.	T
8.	b	8.	F
9.	c	9.	T
10.	b	10.	F

CHAPTER 7

Multiple choice	True/False
1. c	1. T
2. a	2. T
3. b	3. F
4. b	4. F
5. c	5. T
6. e	6. T
7. d	7. F
8. a	8. F
9. a	9. T
10. d	10. F

CHAPTER 8

Multiple choice	True/False
1. a	1. T
2. b	2. F
3. d	3. F
4. e	4. T
5. b	5. F
6. a	6. T
7. a	7. T
8. d	8. T
9. c	9. F
10. a	10. T

CHAPTER 9

Multiple choice	True/False
1. b	1. T
2. b	2. F
3. b	3. F
4. d	4. F
5. e	5. T
6. a	6. T
7. a	7. F
8. b	8. F
9. a	9. T
10. a	10. F

CHAPTER 10

Multiple choice	True/False
1. c	1. F
2. e	2. F
3. b	3. F
4. c	4. T
5. b	5. T
6. b	6. T
7. a	7. F
8. d	8. T
9. a	9. T
10. b	10. F

CHAPTER 11

Multiple choice	True/False
1. c	1. F
2. d	2. T
3. a	3. T
4. b	4. T
5. c	5. T
6. b	6. F
7. d	7. F
8. a	8. T
9. b	9. F
10. d	10. F

CHAPTER 12

Multiple choice	True/False
1. c	1. T
2. c	2. F
3. c	3. F
4. c	4. T
5. a	5. T
6. b	6. F
7. c	7. T
8. b	8. T
9. c	9. F
10. d	10. F

CHAPTER 13

Multiple choice		True/False	
1.	d	1.	T
2.	a	2.	F
3.	e	3.	T
4.	c	4.	T
5.	b	5.	F
6.	b	6.	F
7.	d	7.	T
8.	e	8.	T
9.	b	9.	T
10.	a	10.	T

CHAPTER 16

Multiple choice		True/False	
1.	b	1.	T
2.	c	2.	T
3.	d	3.	T
4.	c	4.	F
5.	b	5.	T
6.	e	6.	F
7.	e	7.	T
8.	a	8.	F
9.	c	9.	T
10.	a	10.	T

CHAPTER 14

Multiple choice		True/False	
1.	c	1.	T
2.	a	2.	F
3.	c	3.	T
4.	c	4.	F
5.	c	5.	T
6.	a	6.	F
7.	d	7.	T
8.	e	8.	T
9.	b	9.	F
10.	b	10.	F

CHAPTER 17

Multiple choice		True/False	
1.	a	1.	F
2.	a	2.	F
3.	d	3.	T
4.	a	4.	T
5.	c	5.	F
6.	a	6.	T
7.	d	7.	T
8.	e	8.	T
9.	e	9.	F
10.	e	10.	T

CHAPTER 15

Multiple choice		True/False	
1.	c	1.	T
2.	a	2.	T
3.	a	3.	F
4.	c	4.	T
5.	a	5.	F
6.	b	6.	T
7.	b	7.	F
8.	c	8.	F
9.	a	9.	T
10.	b	10.	T

CHAPTER 18

Multiple choice		True/False	
1.	d	1.	F
2.	c	2.	T
3.	c	3.	F
4.	a	4.	T
5.	b	5.	F
6.	b	6.	T
7.	a	7.	F
8.	e	8.	T
9.	e	9.	F
10.	b	10.	F

CHAPTER 19

Multiple choice	True/False
1. e	1. T
2. a	2. F
3. d	3. T
4. d	4. T
5. b	5. T
6. b	6. T
7. a	7. F
8. c	8. F
9. b	9. F
10. b	10. T

CHAPTER 21

Multiple choice	True/False
1. a	1. T
2. b	2. F
3. c	3. T
4. a	4. T
5. a	5. T
6. a	6. F
7. c	7. T
8. d	8. T
9. c	9. F
10. b	10. T

CHAPTER 20

Multiple choice	True/False
1. e	1. F
2. d	2. T
3. b	3. T
4. b	4. T
5. b	5. F
6. e	6. T
7. b	7. F
8. e	8. T
9. e	9. T
10. a	10. T

CHAPTER 1

CHAPTER 2

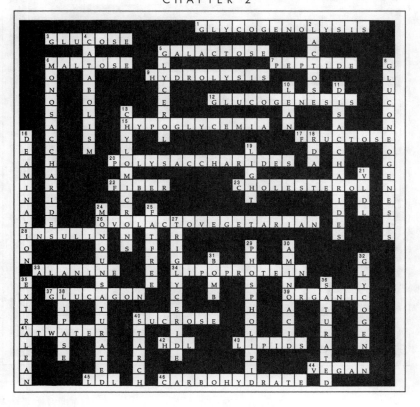

CHAPTER 3

CHAPTER 4

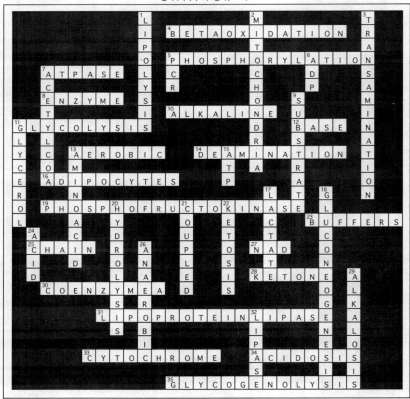

CHAPTER 5

CHAPTER 6

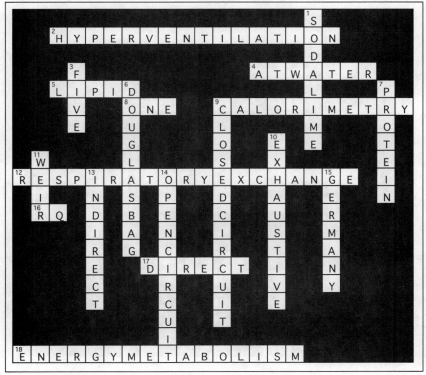

CHAPTER 7

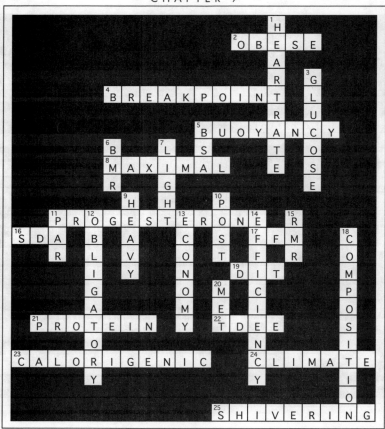

CHAPTER 8

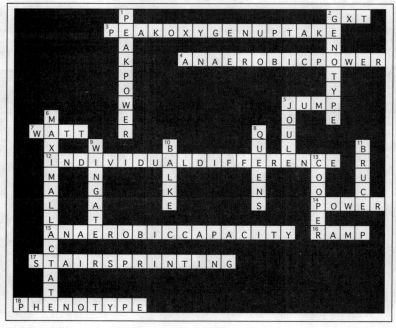

CHAPTER 9

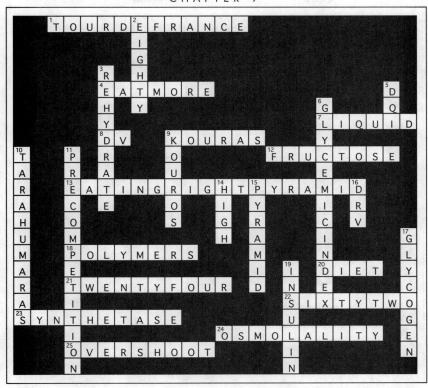

CHAPTER 10

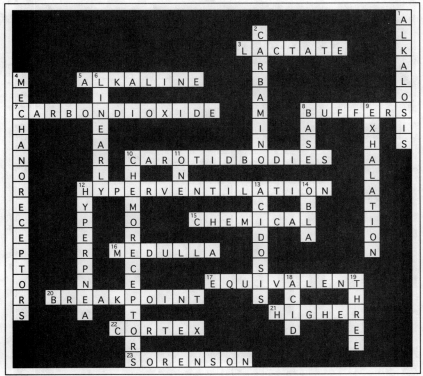

CHAPTER 11

CHAPTER 12

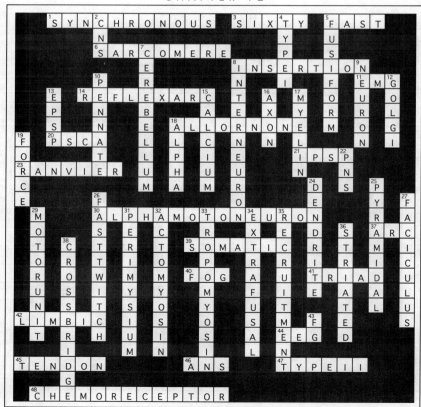

CHAPTER 13

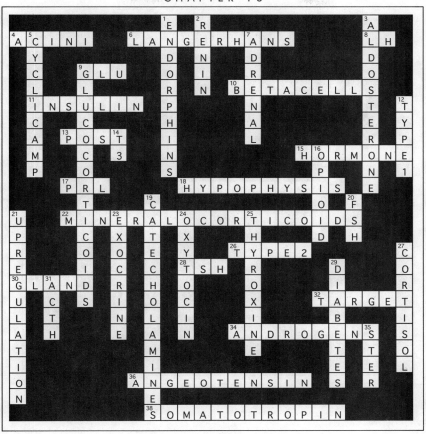

CHAPTER 14

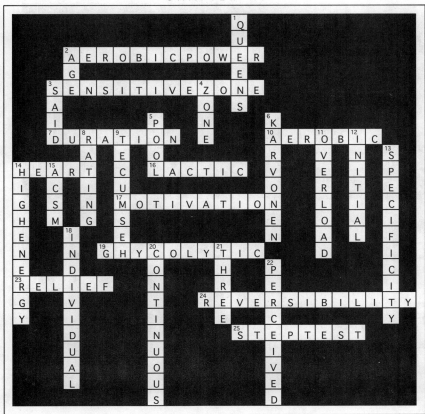

CHAPTER 15

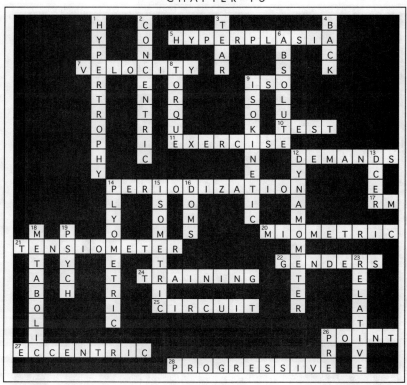

CHAPTER 16

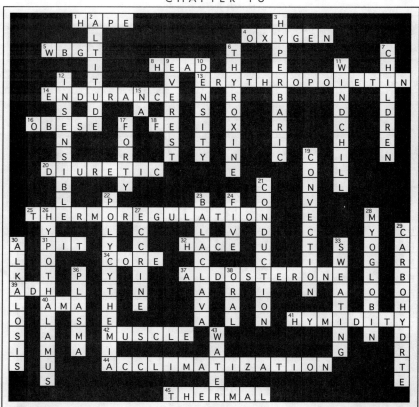

CHAPTER 17

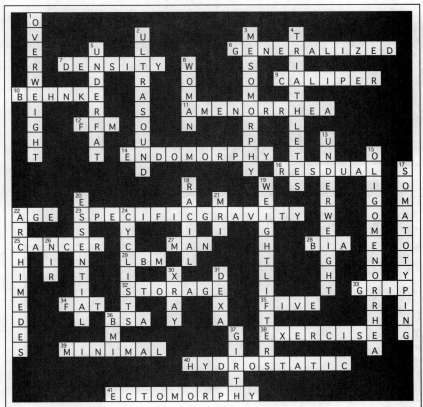

CHAPTER 18

CHAPTER 19

CHAPTER 20

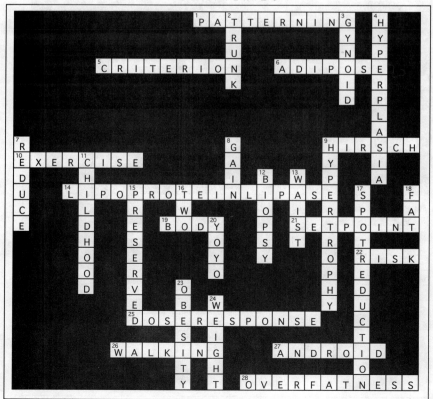

CHAPTER 21

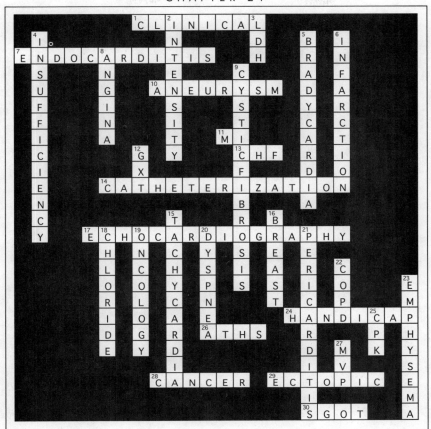

Section IV

■ Appendices

Energy Expenditure in Household, Occupational, Recreational, and Sports Activities[a,b]

■ HOW TO USE APPENDIX A

Refer to the column that comes closest to your present body mass. Multiply the number in this column by the number of minutes you spend in the activity. Suppose that an individual weighing 62.3 kg (137 lb) spends 30 minutes playing a casual game of billiards. To determine the energy cost of participation, multiply the caloric value per minute (2.6 kcal) by 30 to obtain the 30-minute gross expenditure of 78 kcal. If the same individual does aerobic dance for 45 minutes, the **gross** (value includes resting energy expenditure) energy expended would be calculated as 6.4 kcal × 45 minutes, or 288 kcal.

---

[a] All values for energy expenditure are in kilocalories per minute.
[b] Copyright © 1996 by Frank I. Katch, Victor L. Katch, and William D. McArdle, and Fitness Technologies, Inc. P.O. Box 430, Amherst, MA 01004. No part of this appendix may be reproduced in any manner without written permission from the copyright holders.

YOUR BODY WEIGHT

Activity	kg lb	47 104	50 110	53 117	56 123	59 130	62 137	65 143	68 150
Archery		3.1	3.3	3.4	3.6	3.8	4.0	4.2	4.4
Backpacking									
without load		5.7	6.1	6.4	6.8	7.1	7.5	7.9	8.2
with 11 pound load		6.1	6.5	6.8	7.2	7.6	8.0	8.4	8.8
with 22 pound load		6.6	7.0	7.4	7.8	8.3	8.7	9.1	9.5
with 44 pound load		7.0	7.4	7.8	8.2	8.7	9.1	9.6	10.0
Badminton									
leisure		4.6	4.9	5.1	5.4	5.7	6.0	6.3	6.6
tournament		7.0	7.3	7.7	8.1	8.6	9.0	9.4	9.9
Baking, general (F)		1.6	1.8	1.9	2.0	2.1	2.2	2.3	2.4
Baseball									
fielder		2.8	3.0	3.2	3.4	3.6	3.8	4.0	4.1
pitcher		4.2	4.5	4.8	5.0	5.3	5.6	5.9	6.2
Basketball									
competition		7.1	7.4	7.9	8.3	8.7	9.2	9.6	10.1
practice		6.5	6.9	7.3	7.7	8.1	8.6	9.0	9.4
Baton twirling		6.3	6.8	7.3	7.6	8.1	8.5	8.9	9.3
Billiards ("pool")		2.0	2.1	2.2	2.4	2.5	2.6	2.7	2.9
Bookbinding		1.8	1.9	2.0	2.1	2.2	2.4	2.5	2.6
Bowling		4.4	4.8	5.2	5.4	5.7	6.0	6.3	6.6
Boxing									
in ring, match		10.4	11.1	11.8	12.4	13.1	13.8	14.4	15.1
sparring, practice		6.5	6.9	7.3	7.7	8.1	8.6	9.0	9.4
Calisthenics, warm-ups		3.4	3.7	4.0	4.2	4.4	4.7	4.9	5.1
Canoeing									
leisure (2.5 mph)		2.1	2.2	2.3	2.5	2.6	2.7	2.9	3.0
racing ("fast")		4.8	5.2	5.5	5.8	6.1	6.4	6.7	7.0
Car washing		3.3	3.5	3.7	3.9	4.1	4.3	4.5	4.8
Card playing		1.2	1.3	1.3	1.4	1.5	1.6	1.6	1.7
Carpentry, general		2.4	2.6	2.8	2.9	3.1	3.2	3.4	3.5
Carpet sweeping (F)		2.2	2.3	2.4	2.5	2.7	2.8	2.9	3.1
Carpet sweeping (M)		2.3	2.4	2.5	2.7	2.8	3.0	3.1	3.3
Circuit resistance training									
Free weights		4.0	4.3	4.5	4.8	5.0	5.3	5.5	5.8
Hydra-Fitness		6.2	6.6	7.0	7.4	7.8	8.2	8.6	9.0
Nautilus		4.3	4.6	4.9	5.2	5.5	5.8	6.0	6.3
Universal		5.3	5.8	6.2	6.5	6.9	7.2	7.5	7.9
Cleaning (F)		2.9	3.1	3.3	3.5	3.7	3.8	4.0	4.2
Cleaning (M)		2.7	2.9	3.1	3.2	3.4	3.6	3.8	3.9
Coal mining									
drilling coal, rock		4.4	4.7	5.0	5.3	5.5	5.8	6.1	6.4
erecting supports		4.1	4.4	4.7	4.9	5.2	5.5	5.7	6.0
shoveling coal		5.1	5.4	5.7	6.0	6.4	6.7	7.0	7.3
Cooking (F)		2.1	2.3	2.4	2.5	2.7	2.8	2.9	3.1
Cooking (M)		2.3	2.4	2.5	2.7	2.8	3.0	3.1	3.3
Cricket									
batting		3.9	4.2	4.4	4.6	4.9	5.1	5.4	5.6
bowling		4.2	4.5	4.8	5.0	5.3	5.6	5.9	6.1
fielding		3.7	3.9	4.1	4.3	4.8	4.8	5.0	5.3
Croquet		2.8	3.0	3.1	3.3	3.5	3.7	3.8	4.0
Cycling									
leisure, 5.5 mph		3.0	3.2	3.4	3.6	3.8	4.0	4.2	4.4
leisure, 9.4 mph		4.8	5.0	5.3	5.6	5.9	6.2	6.5	6.8
racing, fast		8.0	8.5	9.0	9.5	10.0	10.5	11.0	11.5
Dancing									
aerobic, easy		4.3	4.8	5.2	5.6	5.9	6.2	6.4	6.7
aerobic, medium		4.8	5.2	5.5	5.8	6.1	6.4	6.7	7.0
aerobic, intense		6.3	6.7	7.1	7.5	7.9	8.3	8.7	9.2
ballroom		2.4	2.6	2.7	2.9	3.0	3.2	3.3	3.5
choreographed		5.0	5.2	5.5	5.8	6.1	6.4	6.7	7.0
"twist," "lambada"		8.0	8.4	8.9	9.4	9.9	10.4	10.9	11.4
modern		3.4	3.6	3.8	4.0	4.3	4.5	4.7	4.9

Note: Symbols (M) and (F) denote experiments for males and females, respectively.

71 157	74 163	77 170	80 176	83 183	86 190	89 196	92 203	95 209	98 216
4.6	4.8	5.0	5.2	5.4	5.6	5.8	6.0	6.2	6.4
8.6	9.0	9.3	9.7	10.0	10.4	10.8	11.1	11.5	11.9
9.2	9.5	9.9	10.3	10.7	11.1	11.5	11.9	12.3	12.6
9.9	10.4	10.8	11.2	11.6	12.0	12.5	12.9	13.3	13.7
10.4	10.9	11.3	11.8	12.2	12.6	13.1	13.5	14.0	14.4
6.9	7.2	7.5	7.8	8.1	8.3	8.6	8.9	9.2	9.5
10.4	10.8	11.2	11.6	12.1	12.5	12.9	13.4	13.8	14.3
2.5	2.6	2.7	2.8	2.9	3.0	3.1	3.2	3.3	3.4
4.3	4.5	4.7	4.9	5.1	5.2	5.4	5.6	5.8	6.0
6.4	6.7	7.0	7.2	7.5	7.8	8.0	8.3	8.6	8.9
10.5	10.9	11.4	11.8	12.3	12.7	13.1	13.6	14.0	14.5
9.8	10.2	10.6	11.0	11.5	11.9	12.3	12.7	13.1	13.5
9.5	9.7	9.9	10.1	10.4	10.6	10.8	11.0	11.2	11.4
3.0	3.1	3.2	3.4	3.5	3.6	3.7	3.9	4.0	4.1
2.7	2.8	2.9	3.0	3.2	3.3	3.4	3.5	3.6	3.7
6.9	7.2	7.5	7.7	8.1	8.4	8.6	8.9	9.2	9.5
15.8	16.4	17.1	17.8	18.4	19.1	19.8	20.4	21.1	21.8
9.8	10.2	10.6	11.0	11.5	11.9	12.3	12.7	13.1	13.5
5.3	5.5	5.8	6.0	6.2	6.5	6.7	6.9	7.1	7.3
3.1	3.3	3.4	3.5	3.7	3.8	3.9	4.0	4.2	4.3
7.3	7.6	7.9	8.2	8.5	8.9	9.2	9.5	9.8	10.1
5.0	5.2	5.5	5.7	5.7	5.9	6.1	6.3	6.5	6.9
1.8	1.9	1.9	2.0	2.1	2.2	2.2	2.3	2.4	2.5
3.7	3.8	4.0	4.2	4.3	4.5	4.6	4.8	4.9	5.1
3.2	3.3	3.5	3.6	3.7	3.9	4.0	4.1	4.3	4.4
3.4	3.6	3.7	3.8	4.0	4.1	4.3	4.4	4.6	4.7
6.1	6.3	6.6	6.8	7.1	7.4	7.6	7.9	8.1	8.4
9.4	9.7	10.2	10.5	10.9	11.4	11.7	12.1	12.5	12.9
6.6	6.8	7.1	7.4	7.7	8.0	8.2	8.5	8.8	9.1
8.3	8.6	8.9	9.3	9.6	10.0	10.3	10.7	11.0	11.4
4.4	4.6	4.8	5.0	5.1	5.3	5.5	5.7	5.9	6.1
4.1	4.3	4.5	4.6	4.8	5.0	5.2	5.3	5.5	5.7
6.7	7.0	7.2	7.5	7.8	8.1	8.4	8.6	8.9	9.2
6.2	6.5	6.8	7.0	7.3	7.6	7.8	8.1	8.4	8.6
7.7	8.0	8.3	8.6	9.0	9.3	9.6	9.9	10.3	10.6
3.2	3.3	3.5	3.6	3.7	3.9	4.0	4.1	4.3	4.4
3.4	3.6	3.7	3.8	4.0	4.1	4.3	4.4	4.6	4.7
5.9	6.1	6.4	6.6	6.9	7.1	7.4	7.6	7.9	8.1
6.4	6.7	6.9	7.2	7.5	7.7	8.0	8.3	8.6	8.8
5.6	5.9	6.2	6.5	6.8	7.1	7.4	7.7	8.0	8.3
4.2	4.4	4.5	4.7	4.9	5.1	5.3	5.4	5.6	5.8
4.5	4.7	4.9	5.1	5.3	5.5	5.7	5.9	6.1	6.3
7.1	7.4	7.7	8.0	8.3	8.6	8.9	9.2	9.5	9.8
12.0	12.5	13.0	13.5	14.0	14.5	15.0	15.5	16.1	16.6
6.9	7.2	7.5	7.8	8.1	8.4	8.8	9.1	9.4	9.7
7.3	7.6	7.9	8.2	8.5	8.9	9.2	9.5	9.8	10.1
9.6	10.0	10.4	10.8	11.2	11.6	12.0	12.4	12.8	13.2
3.6	3.8	3.9	4.1	4.2	4.4	4.5	4.7	4.8	5.0
7.3	7.6	7.9	8.2	8.5	8.9	9.2	9.5	9.8	10.1
11.9	12.4	12.9	13.4	13.9	14.4	15.0	15.5	16.0	16.5
5.1	5.3	5.6	5.8	6.0	6.2	6.4	6.7	6.9	7.1

YOUR BODY WEIGHT — continued

Activity	kg lb	47 104	50 110	53 117	56 123	59 130	62 137	65 143	68 150
Digging trenches		6.8	7.3	7.7	8.1	8.6	9.0	9.4	9.9
Drawing (standing)		1.7	1.8	1.9	2.0	2.1	2.2	2.3	2.4
Eating (sitting)		1.1	1.2	1.2	1.3	1.4	1.4	1.5	1.6
Electrical work		2.7	2.9	3.1	3.2	3.4	3.6	3.8	3.9
Farming									
barn cleaning		6.3	6.8	7.2	7.6	8.0	8.4	8.8	9.2
driving harvester		1.9	2.0	2.1	2.2	2.4	2.5	2.6	2.7
driving tractor		1.8	1.9	2.0	2.1	2.2	2.3	2.4	2.5
feeding cattle		4.2	4.3	4.5	4.8	5.0	5.3	5.5	5.8
feeding animals		3.1	3.3	3.4	3.6	3.8	4.0	4.2	4.4
forking straw bales		6.7	6.9	7.3	7.7	8.1	8.6	9.0	9.4
milking by hand		2.5	2.7	2.9	3.0	3.2	3.3	3.5	3.7
milking by machine		1.1	1.2	1.2	1.3	1.4	1.4	1.5	1.6
shoveling grain		4.2	4.3	4.5	4.8	5.0	5.3	5.5	5.8
Fencing									
competition		7.2	7.6	8.1	8.5	9.0	9.4	9.9	10.8
practice		3.6	3.9	4.2	4.4	4.6	4.9	5.1	5.3
Field hockey		6.5	6.7	7.1	7.5	7.9	8.3	8.7	9.1
Fishing		3.0	3.1	3.3	3.5	3.7	3.8	4.0	4.2
Food shopping (F)		3.0	3.1	3.3	3.5	3.7	3.8	4.0	4.2
Football, competition		6.2	6.6	7.0	7.4	7.8	8.2	8.6	9.0
Forestry									
ax chopping, fast		14.0	14.9	15.7	16.6	17.5	18.4	19.3	20.2
ax chopping, slow		4.0	4.3	4.5	4.8	5.0	5.3	5.5	5.8
barking trees		5.8	6.2	6.5	6.9	7.3	7.6	8.0	8.4
carrying logs		8.7	9.3	9.9	10.4	11.0	11.5	12.1	12.6
felling trees		6.2	6.6	7.0	7.4	7.8	8.2	8.6	9.0
hoeing		4.2	4.6	4.8	5.1	5.4	5.6	5.9	6.2
planting by hand		5.1	5.5	5.8	6.1	6.4	6.8	7.1	7.4
sawing by hand		5.7	6.1	6.5	6.8	7.2	7.6	7.9	8.3
sawing, power		3.5	3.8	4.0	4.2	4.4	4.7	4.9	5.1
stacking firewood		4.2	4.4	4.7	4.9	5.2	5.5	5.7	6.0
trimming trees		6.1	6.5	6.8	7.2	7.6	8.0	8.4	8.8
weeding		3.4	3.6	3.8	4.0	4.2	4.5	4.7	4.9
Frisbee		4.7	5.0	5.3	5.5	5.9	6.2	6.4	6.8
Furriery		3.9	4.2	4.4	4.6	4.9	5.1	5.4	5.6
Gardening									
digging		5.9	6.3	6.7	7.1	7.4	7.8	8.2	8.6
hedging		3.3	3.9	4.1	4.3	4.5	4.8	5.0	5.2
mowing		5.3	5.6	5.9	6.3	6.6	6.9	7.3	7.6
raking		2.5	2.7	2.9	3.0	3.2	3.3	3.5	3.7
Golf		4.0	4.3	4.5	4.8	5.0	5.3	5.5	5.8
Gymnastics		3.0	3.3	3.5	3.7	3.9	4.1	4.3	4.5
Handball		6.9	7.2	7.7	8.1	8.5	9.0	9.4	9.8
Horse-grooming		6.0	6.4	6.8	7.2	7.6	7.9	8.3	8.7
Horseback riding									
galloping		6.4	6.9	7.3	7.7	8.1	8.5	8.9	9.3
trotting		5.2	5.5	5.8	6.2	6.5	6.8	7.2	7.5
walking		1.9	2.1	2.2	2.3	2.4	2.5	2.7	2.8
Horseshoes		3.3	3.4	3.5	3.7	3.9	4.1	4.3	4.5
Housework									
mopping floors		2.8	3.1	3.3	3.5	3.7	3.8	4.0	4.2
dusting		3.0	3.3	3.4	3.6	3.8	4.0	4.2	4.4
laundry		3.1	3.4	3.5	3.7	3.9	4.1	4.3	4.5
washing windows		3.2	3.5	3.6	3.8	4.0	4.2	4.4	4.6
vacuuming		3.0	3.3	3.4	3.6	3.8	4.0	4.2	4.4
Hunting		4.1	4.4	4.7	4.9	5.2	5.5	5.7	6.0
Ice hockey		7.4	7.7	8.2	8.6	9.1	9.6	10.0	10.5
Ironing clothes		1.6	1.7	1.7	1.8	1.9	2.0	2.1	2.2
Judo		9.2	9.8	10.3	10.9	11.5	12.1	12.7	13.3

71 157	74 163	77 170	80 176	83 183	86 190	89 196	92 203	95 209	98 216
10.3	10.7	11.2	11.6	12.0	12.5	12.9	13.3	13.8	14.2
2.6	2.7	2.8	2.9	3.0	3.1	3.2	3.3	3.4	3.5
1.6	1.7	1.8	1.8	1.9	2.0	2.0	2.1	2.2	2.3
4.1	4.3	4.5	4.6	4.8	5.0	5.2	5.3	5.5	5.7
9.6	10.0	10.4	10.8	11.2	11.6	12.0	12.4	12.8	13.2
2.8	3.0	3.1	3.2	3.3	3.4	3.6	3.7	3.8	3.9
2.6	2.7	2.8	3.0	3.1	3.2	3.3	3.4	3.5	3.6
6.0	6.3	6.5	6.8	7.1	7.3	7.6	7.8	8.1	8.3
4.6	4.8	5.0	5.2	5.4	5.6	5.8	6.0	6.2	6.4
9.8	10.2	10.6	11.0	11.5	11.9	12.3	12.7	13.1	13.5
3.8	4.0	4.2	4.3	4.5	4.6	4.8	5.0	5.1	5.3
1.6	1.7	1.8	1.8	1.9	2.0	2.0	2.1	2.2	2.3
6.0	6.3	6.5	6.8	7.1	7.3	7.6	7.8	8.1	8.3
11.2	11.7	12.1	12.6	13.1	13.5	14.0	14.4	14.9	15.5
5.6	5.8	6.1	6.3	6.5	6.8	7.0	7.2	7.4	7.7
9.5	9.9	10.3	10.7	11.1	11.5	11.9	12.3	12.7	13.1
4.4	4.6	4.8	5.0	5.1	5.3	5.5	5.7	5.9	6.1
4.4	4.6	4.8	5.0	5.1	5.3	5.5	5.7	5.9	6.1
9.4	9.8	10.2	10.6	11.0	11.4	11.7	12.1	12.5	12.9
21.1	22.0	22.9	23.8	24.7	25.5	26.4	27.3	28.2	29.1
6.0	6.3	6.5	6.8	7.1	7.3	7.6	7.8	8.1	8.3
8.7	9.1	9.5	9.8	10.2	10.6	10.9	11.3	11.7	12.1
13.2	13.8	14.3	14.9	15.4	16.0	16.6	17.1	17.7	18.2
9.4	9.8	10.2	10.6	11.0	11.4	11.7	12.1	12.5	12.9
6.5	6.7	7.0	7.3	7.6	7.8	8.1	8.4	8.6	8.9
7.7	8.1	8.4	8.7	9.0	9.4	9.7	10.0	10.4	10.7
8.7	9.0	9.4	9.8	10.1	10.5	10.9	11.2	11.6	12.0
5.3	5.6	5.8	6.0	6.2	6.5	6.7	6.9	7.1	7.4
6.2	6.5	6.8	7.0	7.3	7.6	7.8	8.1	8.4	8.6
9.2	9.5	9.9	10.3	10.7	11.1	11.5	11.9	12.3	12.6
5.1	5.3	5.5	5.8	6.0	6.2	6.4	6.6	6.8	7.1
7.1	7.4	7.7	8.0	8.2	8.5	8.8	9.1	9.4	9.7
5.9	6.1	6.4	6.6	6.9	7.1	7.4	7.6	7.9	8.1
8.9	9.3	9.7	10.1	10.5	10.8	11.2	11.6	12.0	12.3
5.5	5.7	5.9	6.2	6.4	6.6	6.9	7.1	7.3	7.5
8.0	8.3	8.6	9.0	9.3	9.6	10.0	10.3	10.6	11.0
3.8	4.0	4.2	4.3	4.5	4.6	4.8	5.0	5.1	5.3
6.0	6.3	6.5	6.8	7.1	7.3	7.6	7.8	8.1	8.3
4.7	4.9	5.1	5.3	5.5	5.7	5.9	6.1	6.3	6.5
10.3	10.7	11.2	11.5	12.0	12.5	12.9	13.3	13.7	14.2
9.1	9.5	9.9	10.2	10.6	11.0	11.4	11.8	12.2	12.5
9.7	10.1	10.6	11.0	11.4	11.8	12.2	12.6	13.0	13.4
7.8	8.1	8.5	8.8	9.1	9.5	9.8	10.1	10.5	10.8
2.9	3.0	3.2	3.3	3.4	3.5	3.6	3.8	3.9	4.0
4.7	4.9	5.1	5.3	5.5	5.7	5.9	6.1	6.3	6.5
4.4	4.6	4.8	5.0	5.2	5.4	5.6	5.8	6.0	6.2
4.6	4.7	4.9	5.1	5.3	5.5	5.7	5.9	6.1	6.3
4.7	4.9	5.1	5.3	5.5	5.7	5.9	6.1	6.3	6.5
4.8	5.0	5.2	5.4	5.6	5.8	6.0	6.2	6.4	6.6
4.6	4.8	5.0	5.2	5.4	5.6	5.8	6.0	6.2	6.4
6.2	6.5	6.7	7.0	7.2	7.5	7.8	8.0	8.2	8.5
11.0	11.5	12.0	12.5	13.1	13.6	14.1	14.6	15.1	15.7
2.3	2.4	2.5	2.6	2.7	2.8	2.9	3.0	3.1	3.2
13.8	14.4	15.0	15.6	16.2	16.8	17.4	17.9	18.5	19.1

YOUR BODY WEIGHT—continued

Activity	kg lb	47 104	50 110	53 117	56 123	59 130	62 137	65 143	68 150
Jumping rope									
70 per min		7.6	8.1	8.6	9.1	9.6	10.0	10.5	11.0
80 per min		7.7	8.2	8.7	9.2	9.7	10.2	10.7	11.2
125 per min		8.3	8.9	9.4	9.9	10.4	11.0	11.5	12.0
145 per min		9.3	9.9	10.4	11.0	11.6	12.2	12.8	13.4
Karate		9.5	9.8	10.3	10.9	11.5	12.1	12.7	13.3
Kendo		9.3	9.7	10.2	10.8	11.4	12.0	12.6	13.2
Knitting, sewing		1.1	1.1	1.2	1.2	1.3	1.4	1.4	1.5
Lacrosse		7.0	7.4	7.9	8.3	8.7	9.2	9.6	10.1
Locksmith		2.8	2.9	3.0	3.2	3.4	3.5	3.7	3.9
Lying at ease		1.0	1.1	1.2	1.2	1.3	1.4	1.4	1.5
Machine-tooling									
machining		2.3	2.4	2.5	2.7	2.8	3.0	3.1	3.3
operating lathe		2.5	2.6	2.8	2.9	3.1	3.2	3.4	3.5
operating punch press		4.2	4.4	4.7	4.9	5.2	5.5	5.7	6.0
tapping and drilling		3.2	3.3	3.4	3.6	3.8	4.0	4.2	4.4
welding		2.4	2.6	2.8	2.9	3.1	3.2	3.4	3.5
working sheet metal		2.3	2.4	2.5	2.7	2.8	3.0	3.1	3.3
Marching, rapid		6.7	7.1	7.5	8.0	8.4	8.8	9.2	9.7
Mountain climbing		7.4	7.9	8.4	8.9	9.4	9.9	10.3	10.8
Motorcycle riding		6.5	6.9	7.3	7.7	8.1	8.5	8.9	9.3
Music playing									
accordion (sitting)		1.5	1.6	1.7	1.8	1.9	2.0	2.1	2.2
cello (sitting)		2.0	2.1	2.2	2.3	2.4	2.5	2.7	2.8
conducting		1.9	2.0	2.1	2.2	2.3	2.4	2.5	2.7
drums (sitting)		3.1	3.3	3.5	3.7	3.9	4.1	4.3	4.5
flute (sitting)		1.7	1.8	1.9	2.0	2.1	2.2	2.3	2.4
horn (sitting)		1.4	1.5	1.5	1.6	1.7	1.8	1.9	2.0
organ (sitting)		2.6	2.7	2.8	3.0	3.1	3.3	3.4	3.6
piano (sitting)		1.9	2.0	2.1	2.2	2.4	2.5	2.6	2.7
trumpet (standing)		1.5	1.6	1.6	1.7	1.8	1.9	2.0	2.1
violin (sitting)		2.2	2.3	2.4	2.5	2.7	2.8	2.9	3.1
woodwind (sitting)		1.5	1.6	1.7	1.8	1.9	2.0	2.1	2.2
Paddleball		8.5	8.9	9.4	10.0	10.5	11.0	11.6	12.1
Paddle tennis		8.4	8.6	9.1	9.6	10.1	10.7	11.1	11.7
Painting									
inside projects		1.6	1.7	1.8	1.9	2.0	2.1	2.2	2.3
outside projects		3.7	3.9	4.1	4.3	4.5	4.8	5.0	5.2
scraping		3.1	3.2	3.3	3.5	3.7	3.9	4.1	4.3
Planting seedings		3.3	3.5	3.7	3.9	4.1	4.3	4.6	4.8
Plastering		3.7	3.9	4.1	4.4	4.6	4.8	5.1	5.3
Printing press work		1.7	1.8	1.9	2.0	2.1	2.2	2.3	2.4
Racquetball		8.4	8.9	9.4	10.0	10.5	11.0	11.6	12.1
Roller skating, leisure		5.3	5.8	6.2	6.5	6.9	7.3	7.3	8.0
Rope jumping									
110 rpm		6.7	7.1	7.5	7.9	8.4	8.8	9.2	9.7
120 rpm		6.4	6.8	7.3	7.7	8.1	8.5	8.9	9.3
130 rpm		6.0	6.4	6.8	7.1	7.5	7.7	8.3	8.7
Rowing									
machine, moderate		5.7	6.0	6.3	6.7	7.0	7.4	7.7	8.1
machine, race pace		8.6	8.9	9.4	10.0	10.5	11.0	11.6	12.1
skull, leisure		4.7	5.0	5.3	5.5	5.9	6.2	6.4	6.8
skull, race pace		8.7	8.9	9.4	10.0	10.5	11.0	11.6	12.1
Running, cross-country		7.8	8.2	8.6	9.1	9.6	10.1	10.6	11.1
Running, on flat surface									
11 min, 30 s per mile		6.3	6.8	7.2	7.6	8.0	8.4	8.8	9.2
9 min per mile		9.1	9.7	10.2	10.8	11.4	12.0	12.5	13.1
8 min per mile		9.8	10.8	11.3	11.9	12.5	13.1	13.6	14.2
7 min per mile		10.7	12.2	12.7	13.3	13.9	14.5	15.0	15.6
6 min per mile		11.8	13.9	14.4	15.0	15.6	16.2	16.7	17.3
5 min, 30 s per mile		13.6	14.5	15.3	16.2	17.1	17.9	18.8	19.7
Sailing, leisure		2.1	2.2	2.3	2.5	2.6	2.7	2.9	3.0
Scrubbing floors		5.1	5.5	5.8	6.1	6.4	6.8	7.1	7.4
Scuba diving		10.9	11.2	11.5	11.8	12.1	12.4	12.7	13.0

71 157	74 163	77 170	80 176	83 183	86 190	89 196	92 203	95 209	98 216
11.5	12.0	12.5	13.0	13.4	13.9	14.4	14.9	15.4	15.9
11.6	12.1	12.6	13.1	13.6	14.1	14.6	14.6	15.6	16.1
12.6	13.1	13.6	14.2	14.7	15.2	15.8	16.3	16.8	17.3
14.0	14.6	15.2	15.8	16.4	16.9	17.5	18.1	18.7	19.3
13.8	14.4	15.0	15.6	16.2	16.8	17.4	17.9	18.5	19.1
13.7	14.3	14.9	15.5	16.1	16.7	17.3	17.8	18.4	19.0
1.6	1.6	1.7	1.8	1.8	1.9	2.0	2.0	2.1	2.2
10.4	10.7	11.0	11.2	11.5	11.8	12.1	12.4	12.7	13.0
4.0	4.2	4.4	4.6	4.7	4.9	5.1	5.2	5.4	5.6
1.6	1.6	1.7	1.8	1.8	1.9	2.0	2.0	2.1	2.2
3.4	3.6	3.7	3.8	4.0	4.1	4.3	4.4	4.6	4.7
3.7	3.8	4.0	4.2	4.3	4.5	4.6	4.8	4.9	5.1
6.2	6.5	6.8	7.0	7.3	7.6	7.8	8.1	8.4	8.6
4.6	4.8	5.0	5.2	5.4	5.6	5.8	6.0	6.2	6.4
3.7	3.8	4.0	4.2	4.3	4.5	4.6	4.8	4.9	5.1
3.4	3.6	3.7	3.8	4.0	4.1	4.3	4.4	4.6	4.7
10.1	10.5	10.9	11.4	11.8	12.2	12.6	13.1	13.5	13.9
11.3	11.7	12.2	12.7	13.2	13.7	14.1	14.6	15.0	15.6
9.7	10.1	10.5	10.9	11.3	11.7	12.1	12.5	12.9	13.3
2.3	2.4	2.5	2.6	2.7	2.8	2.8	2.9	3.0	3.1
2.9	3.0	3.2	3.3	3.4	3.5	3.6	3.8	3.9	4.0
2.8	2.9	3.0	3.1	3.2	3.4	3.5	3.6	3.7	3.8
4.7	4.9	5.1	5.3	5.5	5.7	5.9	6.1	6.3	6.6
2.5	2.6	2.7	2.8	2.9	3.0	3.1	3.2	3.3	3.4
2.1	2.1	2.2	2.3	2.4	2.5	2.6	2.7	2.8	2.8
3.8	3.9	4.1	4.2	4.4	4.6	4.7	4.9	5.0	5.2
2.8	3.0	3.1	3.2	3.3	3.4	3.6	3.7	3.8	3.9
2.2	2.3	2.4	2.5	2.6	2.7	2.8	2.9	2.9	3.0
3.2	3.3	3.5	3.6	3.7	3.9	4.0	4.1	4.3	4.4
2.3	2.4	2.5	2.6	2.7	2.8	2.8	2.9	3.0	3.1
12.6	13.2	13.7	14.2	14.8	15.3	15.8	16.4	16.9	17.4
12.2	12.7	13.2	13.7	14.2	14.2	15.2	15.8	16.3	16.8
2.4	2.5	2.6	2.7	2.8	2.9	3.0	3.1	3.2	3.3
5.5	5.7	5.9	6.2	6.4	6.6	6.9	7.1	7.3	7.5
4.5	4.7	4.9	5.0	5.2	5.4	5.6	5.8	6.0	6.2
5.0	5.2	5.4	5.6	5.8	6.0	6.2	6.4	6.7	6.9
5.5	5.8	6.0	6.2	6.5	6.7	6.9	7.2	7.4	7.6
2.5	2.6	2.7	2.8	2.9	3.0	3.1	3.2	3.3	3.4
12.6	13.2	13.7	14.2	14.8	15.3	15.8	16.4	16.9	17.4
8.3	8.6	9.0	9.3	9.7	10.1	10.4	10.8	11.1	11.4
10.1	10.5	10.5	11.3	11.8	12.2	12.6	13.1	13.5	13.9
9.8	10.1	10.6	10.9	11.4	11.8	12.2	12.6	13.0	13.4
9.1	9.4	9.8	10.2	10.6	11.0	11.3	11.7	12.1	12.5
8.5	8.9	9.3	9.7	10.1	10.6	11.1	11.6	12.1	12.6
12.6	13.2	13.7	14.2	14.8	15.3	15.8	16.4	16.9	17.4
7.2	7.6	8.0	8.4	8.8	9.2	9.6	10.0	10.4	10.8
12.6	13.2	13.7	14.2	14.8	15.3	15.8	16.4	16.9	17.4
11.6	12.1	12.6	13.0	13.5	14.0	14.5	15.0	15.5	16.0
9.6	10.0	10.5	10.9	11.3	11.7	12.1	12.5	12.9	13.3
13.7	14.3	14.9	15.4	16.0	16.6	17.2	17.8	18.3	18.9
14.8	15.4	16.0	16.5	17.1	17.7	18.3	18.9	19.4	20.0
16.2	16.8	17.4	17.9	18.5	19.1	19.7	20.3	20.8	21.4
17.9	18.5	19.1	19.6	20.2	20.8	21.4	22.0	22.5	23.1
20.5	21.4	22.3	23.1	24.0	24.9	25.7	26.6	27.5	28.3
3.1	3.3	3.4	3.5	3.7	3.8	3.9	4.1	4.2	4.3
7.7	8.1	8.4	8.7	9.0	9.4	9.7	10.0	10.4	10.7
13.3	13.6	13.9	14.2	14.5	14.8	15.1	15.4	15.7	16.0

YOUR BODY WEIGHT — continued

Activity	kg 47 / lb 104	50 / 110	53 / 117	56 / 123	59 / 130	62 / 137	65 / 143	68 / 150
Shoe repair, general	2.2	2.3	2.4	2.5	2.7	2.8	2.9	3.1
Sitting quietly	1.0	1.1	1.1	1.2	1.2	1.3	1.4	1.4
Skateboarding	5.6	5.8	6.2	6.5	6.9	7.2	7.5	7.9
Skiing, hard snow								
level, moderate speed	5.6	6.0	6.3	6.7	7.0	7.4	7.7	8.1
level, walking speed	6.7	7.2	7.6	8.0	8.4	8.9	9.3	9.7
uphill, "fast" speed	12.9	13.7	14.5	15.3	16.2	17.0	17.8	18.6
Skiing, soft snow								
leisure (F)	4.6	4.9	5.2	5.5	5.8	6.1	6.4	6.7
leisure (M)	5.2	5.6	5.9	6.2	6.5	6.9	7.2	7.5
Skindiving								
considerable motion	13.0	13.8	14.6	15.5	16.3	17.1	17.9	18.8
moderate motion	9.7	10.3	10.9	11.5	12.2	12.8	13.4	14.0
Snorkeling	4.3	4.6	4.9	5.2	5.5	5.8	6.0	6.3
Snowshoeing, soft snow	7.8	8.3	8.8	9.3	9.8	10.3	10.8	11.3
Snowmobiling	3.4	3.7	4.0	4.2	4.4	4.7	4.9	5.1
Soccer	6.5	6.8	7.3	7.7	8.1	8.5	8.9	9.3
Softball	3.3	3.5	3.7	3.9	4.1	4.3	4.5	4.7
Squash	10.0	10.6	11.2	11.9	12.5	13.1	13.8	14.4
Standing quietly (M)	1.3	1.4	1.4	1.5	1.6	1.7	1.8	1.8
Steel mill, working in								
fettling	4.3	4.5	4.7	5.0	5.3	5.5	5.8	6.1
forging	4.7	5.0	5.3	5.6	5.9	6.2	6.5	6.8
hand rolling	6.4	6.9	7.3	7.7	8.1	8.5	8.9	9.3
merchant mill rolling	6.8	7.3	7.7	8.1	8.6	9.0	9.4	9.9
removing slag	8.4	8.9	9.4	10.0	10.5	11.0	11.6	12.1
tending furnace	5.9	6.3	6.7	7.1	7.4	7.8	8.2	8.6
tipping molds	4.3	4.6	4.9	5.2	5.4	5.7	6.0	6.3
Surfing	3.9	4.1	4.3	4.5	4.8	5.0	5.3	5.5
Stock clerking	2.5	2.7	2.9	3.0	3.2	3.3	3.5	3.7
Swimming, fitness swims								
back stroke	7.9	8.5	9.0	9.5	10.0	10.5	11.0	11.5
breast stroke	7.6	8.1	8.6	9.1	9.6	10.0	10.5	11.0
butterfly		8.6	9.1	9.6	10.1	10.7	11.1	11.7
crawl, fast	7.3	7.8	8.3	8.7	9.2	9.7	10.1	10.6
crawl, slow	6.0	6.4	6.8	7.2	7.6	7.9	8.3	8.7
side stroke	5.7	6.1	6.5	6.8	7.2	7.6	7.9	8.3
treading, fast	8.0	8.5	9.0	9.5	10.0	10.5	11.1	11.6
treading, normal	2.9	3.1	3.3	3.5	3.7	3.8	4.0	4.2
Table tennis (ping pong)	3.2	3.4	3.6	3.8	4.0	4.2	4.4	4.6
Tailoring								
cutting	2.0	2.1	2.2	2.3	2.4	2.5	2.7	2.8
hand-sewing	1.5	1.6	1.7	1.8	1.9	2.0	2.1	2.2
machine-sewing	2.2	2.3	2.4	2.5	2.7	2.8	2.9	3.1
pressing	2.9	3.1	3.3	3.5	3.7	3.8	4.0	4.2
Tennis								
competition	6.9	7.3	7.8	8.2	8.7	9.1	9.5	9.9
recreational	5.1	5.5	5.8	6.1	6.4	6.8	7.1	7.4
Typing								
electric (computer)	1.3	1.4	1.4	1.5	1.6	1.7	1.8	1.8
manual	1.5	1.6	1.6	1.7	1.8	1.9	2.0	2.1
Volleyball								
competition	5.9	7.3	7.8	8.2	8.7	9.1	9.5	10.0
recreational	2.4	2.5	2.7	2.8	3.0	3.1	3.3	3.4
Walking, leisure outdoors								
asphalt road	3.8	4.0	4.2	4.5	4.7	5.0	5.2	5.4
fields and hillsides	3.9	4.1	4.3	4.6	4.8	5.1	5.3	5.6
grass track	3.8	4.1	4.3	4.5	4.8	5.0	5.3	5.5
plowed field	3.6	3.9	4.1	4.3	4.5	4.8	5.0	5.2
Walking, treadmill level								
2.0 mph	2.4	2.6	2.8	3.0	3.1	3.3	3.4	3.6
2.5 mph	3.0	3.2	3.4	3.6	3.8	4.0	4.2	4.4
3.0 mph	3.6	3.8	4.0	4.2	4.4	4.6	4.8	5.0
3.5 mph	4.0	4.3	4.6	4.8	5.1	5.3	5.6	6.1
4.0 mph	4.6	4.9	5.2	5.4	5.7	6.0	6.3	6.6

71 157	74 163	77 170	80 176	83 183	86 190	89 196	92 203	95 209	98 216
3.2	3.3	3.5	3.6	3.7	3.9	4.0	4.1	4.3	4.4
1.5	1.6	1.6	1.7	1.7	1.8	1.9	1.9	2.0	2.1
8.3	8.6	8.9	9.3	9.6	10.0	10.3	10.7	11.0	11.4
8.4	8.8	9.2	9.5	9.9	10.2	10.6	10.9	11.3	11.7
10.2	10.6	11.0	11.4	11.9	12.3	12.7	13.2	13.6	14.0
19.5	20.3	21.1	21.9	22.7	23.6	24.4	25.2	26.0	26.9
7.0	7.3	7.5	7.8	8.1	8.4	8.7	9.0	9.3	9.6
7.9	8.2	8.5	8.9	9.2	9.5	9.9	10.2	10.5	10.9
19.6	20.4	21.3	22.1	22.9	23.7	24.6	25.4	26.2	27.0
14.6	15.2	15.9	16.5	17.1	17.7	18.3	19.0	19.6	20.2
6.6	6.8	7.1	7.4	7.7	8.0	8.2	8.5	8.8	9.1
11.8	12.3	12.8	13.3	13.8	14.3	14.8	15.3	15.8	16.3
5.3	5.5	5.8	6.0	6.2	6.5	6.7	6.9	7.1	7.3
9.8	10.1	10.6	10.9	11.4	11.8	12.2	12.6	13.0	13.4
4.9	5.1	5.3	5.5	5.7	5.9	6.1	6.3	6.5	6.7
15.1	15.7	16.3	17.0	17.6	18.2	18.9	19.5	20.1	20.8
1.9	2.0	2.1	2.2	2.2	2.3	2.4	2.5	2.6	2.6
6.3	6.6	6.9	7.1	7.4	7.7	7.9	8.2	8.5	8.7
7.1	7.4	7.7	8.0	8.3	8.6	8.9	9.2	9.5	9.8
9.7	10.1	10.6	11.0	11.4	11.8	12.2	12.6	13.0	13.4
10.3	10.7	11.2	11.6	12.0	12.5	12.9	13.3	13.8	14.2
12.6	13.2	13.7	14.2	14.8	15.3	15.8	16.4	16.9	17.4
8.9	9.3	9.7	10.1	10.5	10.8	11.2	11.6	12.0	12.3
6.5	6.8	7.1	7.4	7.6	7.9	8.2	8.5	8.7	9.0
5.7	6.0	6.3	6.5	6.8	7.0	7.2	7.4	7.6	7.9
3.8	4.0	4.2	4.3	4.5	4.6	4.8	5.0	5.1	5.3
12.0	12.5	13.0	13.5	14.0	14.5	15.0	15.5	16.1	16.6
11.5	12.0	12.5	13.0	13.4	13.9	14.4	14.9	15.4	15.9
12.2	12.7	13.2	13.7	14.2	14.2	15.2	15.8	16.3	16.8
11.1	11.5	12.0	12.5	12.9	13.4	13.9	14.4	14.8	15.3
9.1	9.5	9.9	10.2	10.6	11.0	11.4	11.8	12.2	12.5
8.7	9.0	9.4	9.8	10.1	10.5	10.9	11.2	11.6	12.0
12.1	12.6	13.1	13.6	14.1	14.6	15.1	15.6	16.2	16.7
4.4	4.6	4.8	5.0	5.1	5.3	5.5	5.7	5.9	6.1
4.8	5.0	5.2	5.4	5.6	5.8	6.1	6.3	6.5	6.7
2.9	3.0	3.2	3.3	3.4	3.5	3.6	3.8	3.9	4.0
2.3	2.4	2.5	2.6	2.7	2.8	2.8	2.9	3.0	3.1
3.2	3.3	3.5	3.6	3.7	3.9	4.0	4.1	4.3	4.4
4.4	4.6	4.8	5.0	5.1	5.3	5.5	5.7	5.9	6.1
10.2	10.6	11.1	11.5	11.9	12.4	12.8	13.2	13.7	14.1
7.7	8.1	8.4	8.7	9.0	9.4	9.7	10.0	10.4	10.7
1.9	2.0	2.1	2.2	2.2	2.3	2.4	2.5	2.6	2.6
2.2	2.3	2.4	2.5	2.6	2.7	2.8	2.9	2.9	3.0
3.6	3.7	3.9	4.0	4.2	4.3	4.5	4.6	4.8	4.9
10.5	10.9	11.4	11.8	12.3	12.7	13.1	13.6	14.0	14.5
5.7	5.9	6.2	6.4	6.6	6.9	7.1	7.4	7.6	7.8
5.8	6.1	6.3	6.6	6.8	7.1	7.3	7.5	7.8	8.0
5.8	6.0	6.2	6.5	6.7	7.0	7.2	7.5	7.7	7.9
5.5	5.7	5.9	6.2	6.4	6.6	6.9	7.1	7.3	7.5
3.7	3.9	4.1	4.2	4.4	4.5	4.7	4.9	5.0	5.2
4.5	4.7	4.9	5.1	5.3	5.5	5.7	5.9	6.1	6.3
5.3	5.5	5.7	5.9	6.2	6.5	6.7	6.9	7.1	7.3
6.1	6.4	6.6	6.9	7.1	7.4	7.7	7.9	8.2	8.4
6.9	7.2	7.5	7.8	8.1	8.4	8.7	8.9	9.2	9.5

YOUR BODY WEIGHT—continued

Activity	kg lb	47 104	50 110	53 117	56 123	59 130	62 137	65 143	68 150
Wallpapering		2.3	2.4	2.5	2.7	2.8	3.0	3.1	3.3
Water polo, recreation		7.0	7.4	7.7	8.1	8.5	8.9	9.3	9.7
Water polo, competition		9.4	9.9	10.4	11.0	11.5	12.0	12.5	13.1
Water-skiing		5.6	6.0	6.4	6.7	7.1	7.5	7.8	8.2
Watch repairing		1.2	1.3	1.3	1.4	1.5	1.6	1.6	1.7
Whitewater rafting, recreational		4.1	4.4	4.6	4.9	5.2	5.4	5.7	6.0
Window cleaning		2.9	3.0	3.1	3.3	3.5	3.7	3.8	4.0
Wind surfing		3.3	3.5	3.7	3.9	4.1	4.3	4.6	4.8
Wrestling, competition		9.1	9.7	10.3	10.8	11.4	12.0	12.6	13.2
Writing (sitting)		1.4	1.5	1.5	1.6	1.7	1.8	1.9	2.0
Yoga		2.9	3.1	3.3	3.5	3.7	3.8	4.0	4.2

71 157	74 163	77 170	80 176	83 183	86 190	89 196	92 203	95 209	98 216
3.4	3.6	3.7	3.8	4.0	4.1	4.3	4.4	4.6	4.7
10.1	10.5	10.9	11.3	11.7	12.1	12.5	12.9	13.3	13.7
13.6	14.1	14.7	15.2	15.7	16.3	16.8	17.3	17.9	18.4
8.7	9.1	9.4	9.8	10.1	10.5	10.9	11.2	11.6	12.0
1.8	1.9	1.9	2.0	2.1	2.2	2.2	2.3	2.4	2.5
6.2	6.5	6.7	7.0	7.3	7.5	7.8	8.1	8.3	8.6
4.2	4.4	4.5	4.7	4.9	5.1	5.3	5.4	5.6	5.8
5.0	5.2	5.4	5.6	5.8	6.0	6.2	6.4	6.7	6.9
13.8	14.3	14.9	15.5	16.1	16.7	17.2	17.8	18.4	19.0
2.1	2.1	2.2	2.3	2.4	2.5	2.6	2.7	2.8	2.8
4.4	4.6	4.8	5.0	5.1	5.3	5.5	5.7	5.9	6.1

B

Nutritive Values for Common Foods, Alcoholic and Nonalcoholic Beverages, and Specialty and Fast-Food Items

This appendix has three parts. Part 1 lists nutritive values for common foods, Part 2 lists nutritive values for alcoholic and nonalcoholic beverages, and Part 3 presents nutritive values for specialty and fast-food items. The nutritive values of foods and alcoholic and nonalcoholic beverages are expressed in 1-ounce (28.4 g) portions so comparisons can readily be made between the different food categories. Thus, for example, the protein content of 1.55 g for 1 ounce of banana nut bread can be compared directly to the protein content of 6.28 g for 1 ounce of processed American cheese.

Part 1
Nutritive Values for Common Foods[a]

The foods are grouped into categories and are listed in alphabetical order within each category. The categories include breads, cakes and pies, cookies, candy bars, chocolate, desserts, cereals, cheese, fish, fruits, meats, eggs, dairy products, vegetables, and typical salad bar entries. An additional section labeled Variety consists of food items such as soups, sandwiches, salad dressings, oils, some condiments, and other "goodies." The nutritive value for each food is expressed per ounce or 28.4 g of that food item. The specific values for each food include the caloric content (kcal) for 1 ounce and the protein, total fat, carbohydrate, calcium, iron, vitamin B1, vitamin B2, fiber, and cholesterol content.

---

[a]The information about the nutritive value of the foods was taken from a variety of sources. This includes primarily data from Watt, B.K., and Merrill, A.L.: *Composition of Foods—Raw, Processed and Prepared.* U.S. Department of Agriculture, Washington, DC, 1963; Adams, C, and Richardson, M.: *Nutritive Value of Foods.* Home and Garden Bulletin No. 72, rev, Washington, DC, U.S. Government Printing Office, 1981; and Pennington, J.A.T., and Church, H.N.: *Food Values of Portions Commonly Used*, 14th ed. New York, Harper & Row, 1985. Other sources include a comprehensive database on the Cyber mainframe computer at the University of Massachusetts, the consumer relations departments of manufacturers, and journal articles that evaluated specific foods items. *NA indicates data not available.*

BREADS

	kcal	Protein (g)	Fat (g)	CHO (g)	Ca (mg)	Fe (mg)	B₁ (mg)	B₂ (mg)	Fiber (g)	Cholesterol (mg)
Breads										
Banana nut	91	1.55	4.00	12.7	10.0	0.470	0.054	0.046	0.66	18.3
Boston brown—canned	60	1.26	0.39	13.2	25.8	0.567	0.038	0.025	1.34	1.9
Cornmeal muffin—recipe	91	1.89	3.15	13.2	41.6	0.567	0.069	0.069	1.00	14.5
Croutons—dry	105	3.69	1.04	20.5	35.0	1.020	0.099	0.099	0.09	0
Cracked wheat	74	2.63	0.99	14.2	18.1	0.755	0.108	0.108	1.50	0
Cracked wheat—toast	88	3.00	1.17	16.9	21.6	0.899	0.100	0.128	1.82	0
French—chunk	81	2.67	1.10	14.3	31.6	0.875	0.130	0.097	0.57	0
Italian	78	2.55	0.25	16.0	4.7	0.756	0.116	0.066	0.47	0
Mixed grain	74	2.27	1.05	13.6	30.6	0.907	0.113	0.113	1.78	0
Mixed grain—toast	80	2.47	1.15	14.8	33.3	0.986	0.099	0.123	1.97	0
Oatmeal	74	2.37	1.25	13.6	17.0	0.794	0.130	0.075	1.10	0
Oatmeal—toast	80	2.47	1.36	14.8	18.5	0.863	0.110	0.081	1.20	0
Pita pocket	78	2.94	0.42	15.6	23.2	0.685	0.129	0.061	0.45	0
Pumpernickel	71	2.60	0.98	13.6	20.4	0.777	0.097	0.147	1.67	0
Pumpernickel—toast	78	2.86	1.08	15.0	22.5	0.857	0.088	0.162	1.87	0
Raisin—	77	2.15	1.12	15.0	28.4	0.879	0.093	0.176	0.68	0
Raisin—toasted	92	2.57	1.34	17.6	33.8	1.080	0.081	0.209	0.81	0
Rye—light	74	2.40	1.04	13.6	22.7	0.771	0.116	0.090	1.87	0
Rye—light—toast	84	2.73	1.18	15.5	25.8	0.876	0.107	0.103	2.15	0
Vienna	79	2.72	1.10	14.4	31.2	0.873	0.130	0.100	0.91	0
White	76	2.35	1.10	13.8	35.7	0.806	0.133	0.088	0.54	0
White—toast	84	2.67	1.26	15.7	40.6	0.915	0.121	0.103	0.64	0
Whole wheat	69	2.84	1.22	12.9	20.2	0.964	0.100	0.059	2.10	0
Whole wheat—toasted	79	3.42	1.47	14.4	22.5	1.090	0.090	0.066	2.74	0
Bread crumbs—dry grated	111	3.69	1.42	20.7	34.6	1.160	0.099	0.099	1.15	1.4
Bread crumbs—soft	76	2.35	1.10	13.9	35.9	0.806	0.134	0.088	0.54	0
Bread sticks wo/salt	109	3.40	0.82	21.3	7.9	0.255	0.017	0.020	0.43	0
Bread sticks w/salt	86	2.67	0.89	16.4	13.0	0.243	0.016	0.024	0.41	0

CAKES AND PIES

	kcal	Protein (g)	Fat (g)	CHO (g)	Ca (mg)	Fe (mg)	B₁ (mg)	B₂ (mg)	Fiber (g)	Cholesterol (mg)
Cakes										
Angel food cake	67	1.71	0.09	15.2	23.5	0.123	0.014	0.057	0	0
Boston cream pie	61	0.59	1.89	10.4	6.1	0.142	0.002	0.043	0	4.7
Carrot cake	103	1.05	5.32	13.2	6.9	0.304	0.030	0.035	0	14.6
Cheesecake	86	1.54	5.45	8.1	15.9	0.136	0.009	0.037	0	52.4
Choc cupcake/choc frosting	97	1.24	3.29	16.5	16.9	0.575	0.029	0.041	0	15.2
Coffee cake	91	1.78	2.70	14.8	17.3	0.480	0.054	0.059	0	18.5
Dark fruitcake	109	1.32	4.62	16.5	27.0	0.791	0.053	0.053	0	13.2
Gingerbread cake	91	1.15	2.86	15.2	12.2	0.706	0.042	0.038	0	7.8
Pound cake	113	1.89	4.72	14.2	18.9	0.472	0.047	0.057	0	30.2
Sheet cake—plain	104	1.32	3.96	15.8	18.1	0.429	0.046	0.049	0	20.1
Sheet cake—white frosting	104	0.94	3.28	18.0	14.3	0.281	0.030	0.037	0	16.4
Sponge cake	83	2.01	1.27	16.0	10.8	0.524	0.043	0.046	0	58.8
White cake/coconut	109	1.30	4.05	17.0	13.7	0.454	0.041	0.053	0	1.2
White cake/white frosting	104	1.20	3.59	16.8	13.2	0.399	0.080	0.052	0	1.2
Yellow cake/chocolate frosting	101	1.03	4.48	15.9	9.5	0.509	0.020	0.058	0	15.6
Pies										
Apple pie	73	0.66	3.14	10.7	5.0	0.300	0.031	0.023	0	0
Apple pie—fried	85	0.73	4.67	10.7	4.0	0.312	0.030	0.020	0	4.7
Banana cream pie	46	0.90	1.85	6.7	21.0	0.156	0.022	0.042	0	2.2
Blueberry pie	68	0.72	3.05	9.9	4.7	0.377	0.031	0.025	0	0
Boston cream pie	61	0.59	1.89	10.4	6.1	0.142	0.002	0.043	0	4.7
Cherry pie	74	0.77	3.19	10.9	6.6	0.569	0.034	0.025	0	0
Cherry pie—fried	83	0.68	4.74	10.7	3.7	0.233	0.020	0.020	0	4.3
Chocolate cream pie	50	1.20	2.04	6.9	25.9	0.175	0.024	0.049	0	2.4
Coconut cream pie	57	1.03	2.79	7.2	24.0	0.198	0.021	0.042	0	2.5
Coconut custard pie	66	1.69	3.85	6.3	25.0	0.304	0.029	0.055	0	31.4
Cream pie	85	0.56	4.29	11.0	8.6	0.205	0.011	0.028	0	1.5

CAKES AND PIES — continued

Pies — cont'd	kcal	Protein (g)	Fat (g)	CHO (g)	Ca (mg)	Fe (mg)	B₁ (mg)	B₂ (mg)	Fiber (g)	Cholesterol (mg)
Custard pie	55	1.43	2.65	6.3	23.1	0.269	0.026	0.050	0	27.6
Lemon meringue pie	72	0.95	2.90	10.7	5.1	0.283	0.020	0.028	0	27.7
Mincemeat pie	70	0.65	2.13	12.8	6.9	0.360	0.028	0.024	0	0
Peach pie	73	0.63	3.14	10.9	4.8	0.340	0.031	0.028	0	0
Pecan pie	120	1.30	4.87	18.9	7.2	0.380	0.045	0.034	0	28.1
Pumpkin pie	52	1.28	2.23	7.3	30.0	0.373	0.019	0.042	0	15.5
Strawberry chiffon pie	65	0.85	3.46	8.0	7.7	0.254	0.022	0.023	0	7.1

COOKIES

	kcal	Protein (g)	Fat (g)	CHO (g)	Ca (mg)	Fe (mg)	B₁ (mg)	B₂ (mg)	Fiber (g)	Cholesterol (mg)
Animal cookies	120	1.90	2.89	22.0	3.0	0.918	0.080	0.130	0	0.1
Brownies w/nuts	135	1.84	8.93	15.6	12.8	0.567	0.070	0.070	0	25.5
Butter cookies	130	1.76	4.82	20.2	36.3	0.170	0.011	0.017	0	4.1
Fig bars	106	1.02	1.93	21.4	20.2	0.689	0.039	0.037	0	13.7
Lady fingers	102	2.19	2.19	18.3	11.6	0.515	0.019	0.039	0	101.0
Oatmeal raisin cookies	134	1.64	5.45	19.6	9.8	0.600	0.049	0.044	0	1.1
Peanut butter cookies	145	2.36	8.27	16.5	12.4	0.650	0.041	0.041	0	13.0
Sandwich type cookies	138	1.42	5.67	20.6	8.5	0.992	0.064	0.050	0	0
Shortbread cookies	137	1.77	7.09	17.7	11.5	0.709	0.089	0.080	0	23.9
Sugar cookies	139	1.18	7.09	18.3	29.5	0.532	0.053	0.035	0	17.1
Vanilla wafers	131	1.42	4.96	20.6	11.3	0.567	0.050	0.070	0	17.7

CANDY BARS

	kcal	Protein (g)	Fat (g)	CHO (g)	Ca (mg)	Fe (mg)	B₁ (mg)	B₂ (mg)	Fiber (g)	Cholesterol (mg)
Almond Joy	151	1.69	7.82	18.5	2.0	0.778	0	0	0	0
Sugar-coated almonds	146	3.10	9.12	14.6	39.6	0.775	0.042	0.156	0	0
Bittersweet chocolate	141	1.90	9.73	15.7	13.0	1.040	0.015	0.050	0	0
Caramel — plain or chocolate	115	1.00	2.99	22.0	41.9	0.399	0.010	0.050	0	1.0
Chocolate candy kisses	154	2.10	8.98	15.9	52.9	0.499	0.020	0.080	0	0
Chocolate-coated almonds	161	3.92	12.70	8.0	47.8	1.090	0.052	0.186	0	0
Chocolate-covered coconut	133	0.91	7.10	17.5	8.4	0.614	0.008	0.016	0	0
Chocolate-covered mints	116	0.50	2.99	23.0	16.0	0.299	0.010	0.020	0	0
Chocolate-covered peanuts	159	5.00	11.70	9.8	32.9	0.689	0.086	0.043	0	0
Chocolate-covered raisins	111	1.06	2.71	20.6	12.2	0.663	0.034	0.025	0	0
Chocolate fudge	115	0.56	2.78	21.0	22.0	0.299	0.010	0.030	0	1.0
Chocolate fudge with nuts	114	1.06	4.99	18.8	22.0	0.299	0.016	0.030	0	7.4
English toffee	195	0.89	16.90	9.8	0	0.177	0.470	0.044	0	0
Gum drops	98	0	0.20	24.8	2.0	0.100	0	0	0	0
Hard candy	109	0	0	27.6	6.0	0.100	0	0	0	0
Jelly beans	104	0	0.10	26.4	1.0	0.299	0	0	0	0
Kit Kat	138	1.98	7.25	16.5	42.9	0.369	0.020	0.073	0	0
Krackle	149	2.00	8.09	16.9	50.0	0.400	0.017	0.075	0	0
Malted milk balls	135	2.30	6.99	17.8	62.9	0	0	0	0	0
M&M's plain chocolate	140	1.95	6.08	19.5	46.7	0.449	0.015	0.073	0	0
M&M's peanut chocolate	144	3.23	7.25	16.5	35.4	0.402	0.016	0.056	0	0
Mars bar	136	2.27	6.24	17.0	48.2	0.312	0.014	0.093	0	0
Milk chocolate — plain	145	2.00	8.98	16.0	49.9	0.399	0.020	0.100	0	6.0
Milk chocolate w/almonds	150	2.90	10.40	15.0	60.9	0.559	0.030	0.130	0	4.5
Milk chocolate w/peanuts	155	4.89	11.70	10.0	31.9	0.679	0.112	0.065	0	3.0
Milk chocolate + rice cereal	140	2.00	6.99	18.0	47.9	0.200	0.010	0.080	0	6.0

CANDY BARS—continued

	kcal	Protein (g)	Fat (g)	CHO (g)	Ca (mg)	Fe (mg)	B₁ (mg)	B₂ (mg)	Fiber (g)	Cholesterol (mg)
Milky Way	123	1.53	4.25	20.3	40.6	0.232	0.013	0.070	0	6.6
Mr. Goodbar	151	3.62	9.05	13.9	39.2	0.567	0.030	0.072	0	4.2
Reese's peanut butter cup	151	3.65	9.07	13.9	21.7	0.430	0.020	0.032	0	1.6
Snickers	134	3.08	6.62	17.0	32.4	0.227	0.013	0.050	0	0
Vanilla fudge	118	0.70	3.15	22.0	29.9	0.030	0.006	0.025	0	10.0
Vanilla fudge with nuts	122	1.00	5.01	18.3	25.0	0.159	0.017	0.026	0	8.5

CHOCOLATE

	kcal	Protein (g)	Fat (g)	CHO (g)	Ca (mg)	Fe (mg)	B₁ (mg)	B₂ (mg)	Fiber (g)	Cholesterol (mg)
Baking chocolate	145	3.49	15.00	7.5	22.0	1.900	0.015	0.099	0	0
Bittersweet chocolate	141	1.90	9.73	15.7	13.0	1.040	0.015	0.050	0	0
Milk chocolate—plain	145	2.00	8.98	16.0	49.9	0.399	0.020	0.100	0	6.0
Semi-sweet chocolate chips	143	1.17	10.20	16.2	8.5	0.967	0.017	0.023	0	0
Dark chocolate—sweet	150	1.00	9.98	16.0	7.0	0.599	0.010	0.040	0	0
Chocolate cupcake/ chocolate frosting	97	1.24	3.29	16.5	16.9	0.575	0.029	0.041	0	15.2
Chocolate candy kisses	154	2.10	8.98	15.9	52.9	0.499	0.020	0.080	0	0
Chocolate chip cookies	122	1.54	5.94	18.9	11.0	0.540	0.068	0.155	0	3.4
Chocolate coated almonds	161	3.92	12.70	8.0	47.8	1.090	0.052	0.186	0	0
Chocolate coated peanuts	159	5.00	11.70	9.8	32.9	0.689	0.086	0.043	0	0
Chocolate covered mints	116	0.50	2.99	23.0	16.0	0.299	0.010	0.020	0	0
Chocolate covered raisins	111	1.06	2.71	20.6	12.2	0.663	0.034	0.025	0	0
Chocolate cream pie	50	1.20	2.04	6.9	25.9	0.175	0.024	0.049	0	2.4
Chocolate fudge	115	0.56	2.78	21.0	22.0	0.299	0.010	0.030	0	1.0
Chocolate fudge with nuts	114	1.06	4.99	18.8	22.0	0.299	0.016	0.030	0	7.4
Cake flour-baked value	103	2.08	0.28	22.4	4.5	1.250	0.154	0.096	0	0
Reese's peanut butter cup	151	3.65	9.07	13.9	21.7	0.430	0.020	0.032	0	1.6
Chocolate pudding/recipe	42	0.88	1.25	7.3	27.3	0.142	0.005	0.039	0	4.3
Chocolate pudding instant	34	0.85	0.818	5.9	28.4	0.065	0.009	0.039	0	3.1

DESSERTS AND BREAKFAST PASTRIES

	kcal	Protein (g)	Fat (g)	CHO (g)	Ca (mg)	Fe (mg)	B₁ (mg)	B₂ (mg)	Fiber (g)	Cholesterol (mg)
Apple brown betty	43	0.30	1.60	7.40	5.4	0.130	0.016	0.012	0	3.8
Apple cobbler	55	0.53	1.74	9.57	8.8	0.206	0.023	0.019	0	0.3
Apple crisp	53	0.33	1.93	9.09	7.4	0.278	0.018	0.013	0	0
Apple dumpling	55	0.32	2.37	8.64	7.1	0.253	0.012	0.013	0	0
Banana nut bread	91	1.60	4.00	12.70	10.0	0.470	0.054	0.046	0	18.3
Bread1raisin pudding	60	1.20	2.47	8.54	27.7	0.304	0.030	0.048	0	24.4
Cheesecake	86	1.50	5.45	8.10	15.9	0.136	0.009	0.037	0	52.4
Cherry cobbler	44	0.53	1.37	7.52	8.5	0.391	0.019	0.021	0	0.3
Cherry & cream cheese torte	79	1.28	3.96	10.00	28.5	0.266	0.015	0.052	0	11.2
Vanilla milkshake	32	0.98	0.841	5.09	34.5	0.026	0.013	0.052	0	3.2
Cream puff w/custard fill	72	1.24	4.54	6.83	16.4	0.276	0.015	0.040	0	58.8
Chocolate eclair w/custard fill	79	1.20	4.43	8.96	18.6	0.258	0.019	0.041	0	50.4
Gelatin salad	17	0.43	0	3.99	0.5	0.024	0.002	0.002	0	0
Peach cobbler	28	0.50	1.35	7.96	7.6	0.197	0.018	0.017	0	0.3
Peach crisp	34	0.31	1.06	6.18	4.8	0.203	0.010	0.010	0	0
Crepe, unfilled	49	2.06	1.32	7.07	25.0	0.475	0.045	0.070	0.21	43.0
Pancakes—plain	63	2.10	2.10	9.45	28.4	0.525	0.063	0.074	0.42	16.8
Croissant	117	2.32	6.02	13.40	10.0	1.040	0.085	0.065	0.54	6.5
Danish pastry—plain	109	1.99	5.97	12.90	29.8	0.547	0.080	0.085	0	24.4
Danish pastry w/fruit	102	1.74	5.67	12.20	7.41	0.567	0.070	0.061	0	24.4
Doughnut—cake type	119	1.33	6.75	13.90	13.0	0.454	0.068	0.068	0	11.3
Doughnut—jelly filled	99	1.48	3.84	13.00	12.2	0.349	0.052	0.044	0	0
Doughnut—yeast-raised	111	1.89	6.28	12.30	8.0	0.661	0.132	0.057	0	9.9
Chocolate pudding	42	0.88	1.25	7.28	27.3	0.142	0.005	0.039	0	4.3
Tapioca pudding	38	1.43	1.44	4.85	29.7	0.120	0.012	0.052	0	27.3
Vanilla pudding	32	0.99	1.10	4.50	33.1	0.089	0.009	0.046	0	4.1

DESSERTS AND BREAKFAST PASTRIES — continued

	kcal	Protein (g)	Fat (g)	CHO (g)	Ca (mg)	Fe (mg)	B$_1$ (mg)	B$_2$ (mg)	Fiber (g)	Cholesterol (mg)
Chocolate pudding — instant	34	0.85	0.82	5.89	28.4	0.065	0.009	0.039	0	3.1
Rice pudding	33	0.86	0.86	5.80	28.6	0.107	0.021	0.039	0	3.2
Butterscotch pudding pop	47	1.19	1.29	7.80	37.8	0.020	0.015	0.055	0	0.5
Chocolate pudding pop	49	1.34	1.34	8.20	42.8	0.179	0.015	0.055	0	0.5
Vanilla pudding pop	46	1.19	1.29	7.80	37.8	0.020	0.015	0.055	0	0.5

CEREALS (WITHOUT MILK)

	kcal	Protein (g)	Fat (g)	CHO (g)	Ca (mg)	Fe (mg)	B$_1$ (mg)	B$_2$ (mg)	Fiber (g)	Cholesterol (mg)
All-Bran	70	3.99	0.50	21.0	23.00	4.49	0.369	0.429	8.490	0
Alpha Bits	111	2.20	0.60	24.6	7.99	1.80	0.399	0.399	0.650	0
Apple Jacks	110	1.50	0.10	25.7	2.99	4.49	0.399	0.399	0.200	0
Bran Buds	73	3.95	0.68	21.6	18.90	4.52	0.371	0.439	7.860	0
Bran Chex	90	2.95	0.81	22.6	16.80	4.51	0.347	0.150	5.200	0
Buc Wheats	110	2.00	1.00	24.0	59.90	8.09	0.674	0.764	2.000	0
C.W. Post — plain	126	2.54	4.44	20.3	13.70	4.50	0.380	0.438	0.643	0
C.W. Post w/raisins	123	2.45	4.05	20.3	14.00	4.51	0.358	0.413	0.660	0
Cap'n Crunch	120	1.46	2.60	22.9	4.60	7.53	0.506	0.544	0.709	0
Cap'n Crunchberries	118	1.46	2.35	23.0	8.91	7.32	0.478	0.543	0.324	0
Cap'n Crunch — peanut butter	125	2.03	3.64	21.5	5.67	7.37	0.486	0.567	0.324	0
Cheerios	110	4.24	1.77	19.4	47.30	4.44	0.394	0.394	3.000	0
Cocoa Krispies	109	1.50	0.39	25.2	4.73	1.81	0.394	0.394	0.354	0
Cocoa Pebbles	117	1.35	1.49	24.7	5.40	1.75	0.405	0.405	0.312	0
Corn Bran	98	1.97	1.02	23.9	32.30	9.60	0.299	0.551	5.390	0
Corn Chex	111	2.00	0.10	24.9	2.99	1.80	0.399	0.070	0.499	0
Corn flakes — Kellogg's	110	2.30	0.09	24.4	1.00	1.80	0.367	0.424	0.594	0
Corn flakes — Post Toasties	110	2.30	0.09	24.4	1.00	0.70	0.367	0.424	0.594	0
Corn grits — enriched yellow dry	105	2.49	0.33	22.5	0.55	1.10	0.182	0.107	3.270	0
Corn grits — enriched ckd	17	0.41	0.06	3.7	0.12	0.18	0.028	0.018	0.527	0
Cracklin' Oat Bran	108	2.60	4.16	19.4	18.90	1.80	0.378	0.425	4.280	0
Cream of Rice	15	0.24	0.01	3.3	0.93	0.05	0.012	0	0.163	0
Cream of Wheat	16	0.42	0.07	3.4	6.27	1.27	0.028	0.008	0.395	0
Crispy Wheat 'n Raisins	99	2.00	0.46	23.1	46.80	4.48	0.396	0.396	1.320	0
Farina — cooked	14	0.41	0.02	3.0	0.49	0.14	0.023	0.015	0.389	0
Fortified Oat Flakes	105	5.32	0.41	20.5	40.20	8.09	0.354	0.413	0.827	0
40% Bran Flakes — Kellogg's	91	3.60	0.54	22.2	13.80	8.14	0.369	0.430	0.850	0
40% Bran Flakes — Post	92	3.20	0.45	22.3	12.70	4.50	0.374	0.435	3.800	0
Froot Loops	111	1.70	1.00	25.0	2.99	4.49	0.399	0.399	0.299	0
Frosted Mini-Wheats	102	2.93	0.27	23.4	9.15	1.83	0.366	0.457	2.160	0
Frosted Rice Krispies	109	1.30	0.10	25.7	1.00	1.80	0.399	0.399	0.998	0
Fruit & Fiber w/apples	90	2.99	1.00	22.0	9.98	4.49	0.374	0.424	4.190	0
Fruit & Fiber w/dates	90	2.99	1.00	21.0	9.98	4.49	0.374	0.424	4.190	0
Fruitful Bran	92	2.50	0	22.5	8.34	6.75	0.313	0.354	4.170	0
Fruity Pebbles	115	1.10	1.50	24.4	2.99	1.80	0.399	0.399	0.226	0
Golden Grahams	109	1.60	1.09	24.1	17.40	4.50	0.363	0.436	1.670	0
Granola — homemade	138	3.49	7.69	15.6	17.70	1.12	0.170	0.072	2.970	0
Granola — Nature Valley	126	2.89	4.92	18.9	17.80	0.95	0.098	0.048	2.960	0
Grape Nuts	100	3.28	0.11	23.2	10.90	1.22	0.398	0.398	1.840	0
Grape Nuts Flakes	102	2.99	0.30	23.2	11.00	4.49	0.399	0.399	1.900	0
Honey & Nut Corn Flakes	113	1.80	1.50	23.3	2.99	1.80	0.399	0.399	0.299	0
Honey Bran	96	2.51	0.57	23.2	13.00	4.54	0.405	0.405	3.160	0
Honey Comb	111	1.68	0.52	25.3	5.15	1.80	0.387	0.387	0.387	0
Honey Nut Cheerios	107	3.09	0.69	22.8	19.80	4.47	0.344	0.430	0.790	0
King Vitamin	115	1.49	1.62	24.0	NA	17.10	0.124	1.430	0.135	0
Kix	109	2.49	0.70	23.3	34.80	8.06	0.398	0.398	0.398	0
Life	104	5.22	0.52	20.3	99.20	7.47	0.612	0.644	0.902	0
Lucky Charms	111	2.57	1.06	23.1	31.90	4.52	0.354	0.443	0.624	0
Malt-O-Meal	14	0.43	0.03	3.1	0.59	1.13	0.057	0.028	0.354	0
Maypo — cooked	1	0.02	0.01	0.1	0.52	0.04	0.003	0.003	0.012	0
Nutri-Grain — barley	106	3.11	0.21	23.4	7.60	1.00	0.346	0.415	1.660	0
Nutri-Grain — corn	108	2.30	0.68	24.0	0.68	0.60	0.338	0.405	1.750	0
Nutri-Grain — rye	102	2.48	0.21	24.0	5.67	0.80	0.354	0.425	2.160	0

CEREALS (WITHOUT MILK)—continued

	kcal	Protein (g)	Fat (g)	CHO (g)	Ca (mg)	Fe (mg)	B₁ (mg)	B₂ (mg)	Fiber (g)	Cholesterol (mg)
Nutri-Grain—wheat	102	2.45	0.32	24.0	7.73	0.80	0.387	0.451	1.800	0
Oatmeal—prepared	18	0.73	0.29	3.1	2.42	0.19	0.032	0.006	0.497	0
Rolled Oats	109	4.55	1.78	19.0	14.70	1.19	0.206	0.038	3.090	0
Instant Oatmeal w/apples	26	0.74	0.30	5.0	30.00	1.15	0.091	0.053	0.552	0
Instant Oatmeal w/bran & raisins	23	0.71	0.28	4.4	25.20	1.10	0.081	0.092	0.480	0
Instant Oatmeal w/maple	30	0.84	0.35	5.8	29.60	1.16	0.097	0.059	0.530	0
Instant Oatmeal w/cinnamon & spice	31	0.85	0.34	6.2	30.30	1.17	0.099	0.60	0.510	0
Instant Oatmeal w/raisins & spice	29	0.77	0.32	5.7	29.60	1.18	0.092	0.065	0.556	0
100% Bran	77	3.57	1.42	20.7	19.80	3.49	0.687	0.773	8.380	0
100% Natural	135	3.02	6.02	18.0	48.90	0.83	0.085	0.150	3.390	0
100% Natural—w/apples	130	2.92	5.32	19.0	42.80	0.79	0.090	0.158	1.300	0
100% Natural—w/raisins & dates	128	2.89	5.23	18.7	41.20	0.80	0.077	0.165	1.080	0
Product 19	108	2.75	0.17	23.5	3.44	18.00	1.460	1.720	0.369	0
Puffed Rice	111	1.79	0.20	25.5	2.03	0.30	0.030	0.028	0.227	0
Puffed Wheat	104	4.25	0.24	22.4	7.09	1.35	0.047	0.070	5.430	0
Quisp	117	1.42	2.08	23.6	8.50	5.96	0.510	0.718	0.378	0
Raisin Bran—Kellogg's	91	3.07	0.46	21.4	14.50	13.90	0.293	0.332	3.410	0
Raisin Bran—Post	86	2.68	0.55	21.4	13.70	4.56	0.373	0.430	3.190	0
Raisins, Rice & Rye	96	1.60	0.06	24.2	6.16	3.45	0.308	0.370	0.308	0
Ralston—cooked	15	0.62	0.09	3.2	1.57	0.18	0.022	0.020	0.370	0
Rice Chex	112	1.49	1.00	25.2	3.88	1.79	0.400	0.298	1.840	0
Rice Krispies	109	1.86	0.20	24.2	3.91	1.76	0.391	0.391	0.312	0
Roman Meal—dry	91	4.07	0.60	20.4	18.40	1.31	0.142	0.069	0.905	0
Roman Meal—cooked	17	0.77	0.11	3.9	3.45	0.25	0.028	0.014	0.877	0
Shredded Wheat	102	3.09	0.71	22.5	11.00	1.20	0.070	0.080	3.100	0
Shredded wheat	97	3.06	0.45	16.4	11.20	0.89	0.082	0.075	2.900	0
Special K	111	5.58	0.10	21.3	7.97	4.48	0.399	0.399	0.266	0
Sugar Corn Pops	108	1.40	0.10	25.6	0.10	1.80	0.399	0.399	0.100	0
Sugar Frosted Flakes	108	1.46	0.08	25.7	0.81	1.78	0.405	0.405	0.446	0
Sugar Smacks	106	2.00	0.50	24.7	2.99	1.80	0.369	0.429	0.319	0
Super Golden Crisp	106	1.80	0.26	25.6	6.01	1.80	0.344	0.430	0.430	0
Team	111	1.82	0.48	24.3	4.05	1.73	0.371	0.425	0.270	0
Total	105	2.84	0.60	22.3	172.00	18.00	1.460	1.720	2.060	0
Trix	108	1.50	0.40	24.9	5.99	4.49	0.399	0.399	0.184	0
Wheat & Raisin Chex	97	2.68	0.21	22.6	NA	4.04	0.263	0.315	1.890	0
Wheat Chex	104	2.77	0.68	23.3	11.00	4.50	0.370	0.105	2.100	0
Wheat germ—toasted	108	8.25	3.09	14.0	12.50	2.19	0.474	0.233	3.910	0
Wheat germ w/brown sugar, honey	107	6.19	2.30	17.2	8.98	1.93	0.349	0.180	3.390	0
Wheatena—cooked	16	0.58	0.13	3.4	1.28	0.16	0.002	0.006	0.385	0
Wheaties	99	2.74	0.51	22.6	43.00	4.50	0.391	0.391	2.540	0
Whole wheat berries	16	0.54	0.11	3.2	1.70	0.17	0.023	0.006	0.680	0
Whole wheat cereal—cooked	18	0.58	0.11	3.9	1.99	0.18	0.020	0.014	0.457	0

CHEESE

	kcal	Protein (g)	Fat (g)	CHO (g)	Ca (mg)	Fe (mg)	B1 (mg)	B2 (mg)	Fiber (g)	Cholesterol (mg)
American—processed	106	6.28	8.84	0.45	174	0.110	0.008	0.111	0	27.0
American cheese food—cold pack	94	5.23	6.78	2.36	145	0.240	0.009	0.274	0	18.0
American cheese spread	82	5.16	6.00	2.48	159	0.090	0.014	0.380	0	16.0
Blue	100	6.09	8.14	0.66	150	0.090	0.008	0.395	0	21.0
Brick	105	6.40	8.40	0.79	191	0.130	0.004	0.159	0	27.0
Brie	95	5.87	7.84	0.13	52	0.140	0.020	0.147	0	28.0
Camembert	85	5.60	6.86	0.13	110	0.094	0.008	0.138	0	20.0
Caraway	107	7.13	8.27	0.87	191	0.100	0.009	0.196	0	25.0
Cheddar	114	7.05	9.38	0.36	204	0.197	0.008	0.106	0	29.9
Cheshire	110	6.60	8.66	1.36	182	0.060	0.013	0.198	0	28.9
Colby	112	6.73	9.08	0.73	194	0.216	0.004	0.171	0	27.0
Cottage	29	3.54	1.20	0.76	17	0.040	0.006	0.115	0	4.2

CHEESE—continued

	kcal	Protein (g)	Fat (g)	CHO (g)	Ca (mg)	Fe (mg)	B₁ (mg)	B₂ (mg)	Fiber (g)	Cholesterol (mg)
Cottage—lowfat 2%	26	3.90	0.55	1.03	20	0.045	0.007	0.052	0	2.4
Cottage—lowfat 1%	21	3.51	0.29	0.77	18	0.040	0.006	0.115	0	1.3
Cottage—dry curd	24	4.89	0.12	0.52	9	0.065	0.007	0.004	0	2.0
Cottage—w/fruit	35	2.80	0.96	3.78	14	0.031	0.005	0.115	0	3.1
Cream	99	2.10	9.87	0.75	23	0.337	0.005	0.056	0	30.9
Edam	101	7.07	7.79	0.40	207	0.125	0.010	0.274	0	25.0
Feta	75	4.49	6.19	1.16	140	0.180	0.040	0.315	0	25.0
Fontina	110	7.25	8.62	0.44	156	0.060	0.006	NA	0	32.9
Gjetost	132	2.74	8.32	12.00	113	0.130	0.009	0.170	0	25.0
Gorgonzola	111	6.99	8.98	0	149	0.120	0.010	0.512	0	25.0
Gouda	101	7.06	7.72	0.63	198	0.070	0.009	0.232	0	31.9
Gruyere	117	8.44	9.05	0.10	286	0.060	0.017	0.095	0	30.9
Liederkranz	87	4.99	7.99	0	110	0.120	0.010	0.389	0	21.0
Limburger	93	5.67	7.59	0.14	141	0.040	0.023	0.227	0	26.0
Monterey jack	106	6.93	8.56	0.19	212	0.200	0.004	0.119	0	26.0
Mozzarella—skim, low moist	80	7.60	4.67	0.89	207	0.076	0.006	0.150	0	15.0
Mozzarella—whole milk, regular	80	5.50	5.75	0.63	147	0.050	0.004	0.106	0	22.0
Mozzarella—whole milk, moist	90	6.10	7.19	0.43	163	0.060	0.005	0.119	0	25.0
Muenster	104	6.40	8.42	0.32	203	0.125	0.004	0.178	0	27.0
Neufchatel	74	2.82	6.70	0.83	21	0.080	0.004	0.113	0	22.0
Parmesan—hard	111	10.00	7.30	0.91	335	0.230	0.010	0.453	0	19.0
Parmesan—grated	129	11.80	8.50	1.06	389	0.270	0.013	0.527	0	22.0
Pimento processed	106	6.26	8.82	0.49	174	0.120	0.008	0.404	0	27.0
Port du salut	100	6.73	7.99	0.16	84	0.140	0.004	0.151	0	34.9
Provolone	100	7.13	7.54	0.61	214	0.146	0.005	0.248	0	20.0
Ricotta—part skim	39	3.23	2.25	1.45	77	0.126	0.006	0.052	0	8.8
Ricotta—whole milk	49	3.19	3.68	0.86	59	0.108	0.004	0.024	0	14.3
Romano	110	9.00	7.63	1.03	301	0.230	0.010	0.339	0	28.9
Romano—grated	128	10.50	8.86	1.20	350	0.270	0.013	0.394	0	32.9
Roquefort	105	6.10	8.93	0.57	188	0.172	0.010	0.512	0	26.0
Swiss	107	8.03	7.79	0.96	272	0.050	0.006	0.074	0	26.0
Swiss processed	95	7.00	6.97	0.60	219	0.170	0.004	0.078	0	24.0

FISH

	kcal	Protein (g)	Fat (g)	CHO (g)	Ca (mg)	Fe (mg)	B₁ (mg)	B₂ (mg)	Fiber (g)	Cholesterol (mg)
Bass—freshwater raw	32	5.36	1.05	0	22.7	0.422	0.028	0.009	0	19.3
Bluefish—baked/broiled	45	7.43	1.42	0	2.6	0.174	0.022	0.030	0	17.9
Bluefish—fried in crumbs	58	6.44	2.78	1.33	2.3	0.151	0.017	0.023	0	17
Bluefish—raw	35	5.67	1.20	0	2.0	0.136	0.016	0.023	0	16.7
Carp—raw	36	5.05	1.59	0	11.6	0.352	0.013	0.011	0	18.7
Catfish—channel—raw	33	5.16	1.20	0	11.3	0.275	0.013	0.030	0	16.4
Cod—baked w/butter	37	6.46	0.94	0	5.7	0.139	0.025	0.022	0	17.0
Cod—batter-fried	56	5.56	2.92	2.13	22.7	0.142	0.011	0.011	0	15.6
Cod—baked/broiled	30	6.46	0.24	0	4.0	0.139	0.025	0.022	0	15.6
Cod—poached	29	6.24	0.24	0	4.0	0.139	0.025	0.022	0	15.6
Cod—steamed	29	6.24	0.24	0	4.0	0.139	0.025	0.022	0	15.9
Cod—smoked	22	5.19	0.17	0	4.0	0.113	0.023	0.020	0	14.2
Cod—Atlantic—raw	23	5.05	0.19	0	4.5	0.108	0.022	0.018	0	12.2
Cod liver oil	255	0	28.40	0	0	0	0	0	0	162.0
Eel—smoked	94	5.27	7.88	0.23	26.9	0.198	0.040	0.099	0	19.8
Haddock—breaded/fried	58	5.67	3.00	2.33	11.3	0.384	0.020	0.033	0.01	18.3
Haddock—smoked	33	7.14	0.27	0	13.9	0.397	0.013	0.014	0	21.8
Haddock—raw	22	5.36	0.20	0	9.4	0.298	0.010	0.010	0	16.2
Herring—pickled	74	4.03	5.10	2.73	21.8	0.346	0.010	0.039	0	3.7
Herring—smoked/ kippered	62	6.97	3.52	0	23.8	0.428	0.036	0.090	0	22.7
Herring—canned w/liquid	59	5.64	3.86	0	41.7	0.879	0.007	0.051	0	27.5
Mackerel—fried	49	7.00	2.35	0	4.3	0.445	0.045	0.116	0	19.8
Mackerel—Atlantic— baked/broiled	74	6.78	5.05	0	4.3	0.445	0.045	0.117	0	21.3
Mackerel—Atlantic—raw	58	5.27	3.94	0	3.4	0.462	0.050	0.088	0	19.8

FISH—continued

	kcal	Protein (g)	Fat (g)	CHO (g)	Ca (mg)	Fe (mg)	B₁ (mg)	B₂ (mg)	Fiber (g)	Cholesterol (mg)
Mackerel—Pacific—raw	45	6.12	2.84	0	2.3	0.567	0.043	0.096	0	22.7
Northern pike—raw	25	5.47	0.20	0	16.2	0.156	0.017	0.018	0	11.0
Ocean perch—breaded/fried	62	5.34	3.67	2.33	30.7	0.400	0.033	0.037	0.03	15.3
Pollock—baked/broiled	28	6.60	0.31	0	19.3	0.149	0.014	0.057	0	19.8
Pollock—poached	36	6.60	0.31	0	17.0	0.149	0.010	0.050	0	19.8
Salmon—broiled/baked	61	7.74	3.10	0	2.0	0.157	0.061	0.048	0	24.7
Coho salmon—steamed/poached	52	7.77	2.14	0	8.22	0.252	0.057	0.031	0	13.9
Smoked salmon—Chinook	33	5.17	1.22	0	03.0	0.240	0.007	0.029	0	6.7
Atlantic salmon—small can	36	5.05	1.62	0	3.1	0.204	0.057	0.097	0	17.0
Pink salmon—raw	33	5.64	0.98	0	11.3	0.218	0.040	0.057	0	14.7
Sardines	59	7.00	3.24	0	108.0	0.826	0.023	0.064	0	40.4
Sea trout steelhead—raw	30	4.73	1.02	0	4.8	0.077	0.023	0.057	0	23.5
Sea trout steelhead—cooked	37	6.07	1.42	0	5.7	0.088	0.024	0.064	0	32.3
Shad—baked with bacon	57	6.58	3.20	0	6.8	0.170	0.037	0.074	0	17.0
Smelt—rainbow—raw	28	4.99	0.69	0	17.0	0.255	0.016	0.034	0	19.8
Snapper—baked or broiled	36	7.46	0.49	0	11.3	0.068	0.015	0.021	0	13.3
Snapper—raw	28	5.81	0.38	0	9.1	0.051	0.013	0.017	0	10.5
Sole/flounder—baked w/butter	40	5.34	2.00	0	5.3	0.093	0.023	0.032	0	22.7
Sole/flounder—baked/broiled	33	6.84	0.43	0	5.3	0.093	0.023	0.032	0	19.3
Sole/flounder—batter-fried	83	4.47	5.10	4.07	16.7	0.239	0.057	0.042	0.01	15.0
Sole/flounder—breaded/fried	53	4.96	2.55	2.54	11.3	0.128	0.037	0.034	0	15.0
Sole/flounder—steamed	26	5.67	0.33	0	4.5	0.079	0.017	0.027	0	14.7
Sole/flounder—raw	26	5.33	0.34	0	5.1	0.102	0.025	0.022	0	13.6
Lemon sole—raw	23	4.85	0.21	0	4.8	0.088	0.026	0.023	0	17.0
Lemon sole—fried w/crumbs	56	4.56	3.12	2.64	26.9	0.176	0.020	0.023	0	18.4
Lemon sole—steamed	26	5.84	0.26	0	6.0	0.147	0.026	0.026	0	17.0
Swordfish—raw	34	5.61	1.14	0	1.1	0.230	0.010	0.027	0	11.0
Swordfish—broiled/baked	44	7.20	1.46	0	1.7	0.295	0.012	0.033	0	14.2
Trout—baked/broiled	43	7.47	1.22	0	24.3	0.690	0.024	0.064	0	20.7
Tuna—oil pack	56	8.26	2.34	0	3.8	0.395	0.010	0.030	0	5.0
Tuna—water pack	37	8.38	0.14	0	3.4	0.409	0.010	0.033	0	16.0
Tuna—raw	31	6.63	0.27	0	4.5	0.207	0.123	0.013	0	12.8
Whiting—flour/bread-fried	54	5.13	1.56	1.98	11.3	0.198	0.023	0.020	0	18.4

FRUITS

	kcal	Protein (g)	Fat (g)	CHO (g)	Ca (mg)	Fe (mg)	B₁ (mg)	B₂ (mg)	Fiber (g)	Cholesterol (mg)
Apple w/peel	16	0.055	0.100	4.31	2.05	0.051	0.005	0.004	0.709	0
Apple slices w/peel—fresh	17	0.054	0.100	4.33	2.06	0.052	0.005	0.004	0.709	0
Apple juice—canned/bottled	13	0.017	0.032	3.32	1.94	0.105	0.006	0.005	0.034	0
Apple juice—frozen concentrate	47	0.144	0.105	11.60	5.78	0.258	0.003	0.015	0.089	0
Applesauce—sweetened	22	0.052	0.052	5.67	1.11	0.111	0.003	0.008	0.397	0
Apricot—fresh halves	14	0.397	0.110	3.15	4.02	0.154	0.009	0.011	0.538	0
Apricot halves—light syrup	18	0.150	0.013	4.67	3.34	0.110	0.005	0.006	0.319	0
Apricot nectar—canned	16	0.104	0.025	4.08	2.03	0.108	0.003	0.004	0.170	0
Avocado—average	46	0.563	0.340	2.10	3.07	0.284	0.030	0.035	2.720	0
Banana—fresh slices	26	0.293	0.136	6.63	1.74	0.088	0.013	0.028	0.578	0
Blackberries—canned	26	0.370	0.040	6.54	5.98	0.184	0.008	0.011	1.011	0
Blackberries—fresh	15	0.205	0.110	3.62	9.06	0.158	0.008	0.011	1.910	0
Blackberries—frozen	18	0.334	0.122	4.45	8.26	0.173	0.008	0.013	1.460	0

FRUITS—continued

	kcal	Protein (g)	Fat (g)	CHO (g)	Ca (mg)	Fe (mg)	B₁ (mg)	B₂ (mg)	Fiber (g)	Cholesterol (mg)
Blueberries—fresh	16	0.190	0.108	4.00	1.76	0.047	0.014	0.014	0.763	0
Blueberries-frozen unsweetened	14	0.119	0.181	3.46	2.19	0.051	0.009	0.010	0.658	0
Boysenberries—frozen	14	0.314	0.075	3.46	7.73	0.240	0.015	0.010	1.100	0
Sour cherries—frozen	13	0.260	0.124	3.13	3.66	0.150	0.012	0.010	0.384	0
Sweet cherries—fresh	20	0.340	0.272	4.69	4.10	0.110	0.014	0.017	0.430	0
Sweet cherries—frozen	25	0.325	0.037	6.34	3.39	0.099	0.008	0.013	0.224	0
Cranberries—whole—raw	14	0.110	0.057	3.58	2.09	0.057	0.009	0.006	1.190	0
Cranberry/apple juice	19	0.015	0.090	4.82	2.02	0.017	0.001	0.006	0.070	0
Cranberry juice cocktail	16	0.009	0.015	4.03	0.90	0.043	0.002	0.002	0.085	0
Date—whole—each	78	0.557	0.127	20.80	9.22	0.342	0.026	0.028	2.300	0
Figs—medium—fresh	21	0.215	0.085	5.44	10.20	0.102	0.017	0.014	1.050	0
Fig—dried—each	72	0.864	0.330	18.50	40.80	0.634	0.020	0.025	3.140	0
Fruit cocktail—heavy syrup	21	0.111	0.020	5.36	1.78	0.081	0.005	0.005	0.280	0
Fruit cocktail—light syrup	16	0.114	0.020	4.23	1.80	0.082	0.005	0.005	0.284	0
Grapefruit half—pink/red	9	0.157	0.028	2.18	3.00	0.034	0.010	0.006	0.369	0
Grapefruit half—white	10	0.195	0.029	2.38	3.36	0.017	0.010	0.006	0.368	0
Grapefuit sections/fresh	9	0.179	0.028	2.29	3.33	0.025	0.010	0.006	0.370	0
Grapefruit sections—canned	17	0.160	0.028	4.38	4.02	0.114	0.010	0.006	0.313	0
Grapefruit juice—fresh	11	0.142	0.029	2.60	2.53	0.056	0.011	0.006	0.113	0
Grapefruit juice—sweetened	13	0.164	0.026	3.15	2.27	0.102	0.011	0.007	0.076	0
Grapefruit juice—unsweetened	11	0.148	0.028	2.54	1.95	0.057	0.012	0.006	0.077	0
Grapefruit juice—frozen concentrate	41	0.548	0.137	9.86	7.67	0.140	0.041	0.022	0.383	0
Grapes—Thompson	20	0.188	0.163	5.03	3.01	0.073	0.026	0.016	0.333	0
Grape juice—bottled/canned	17	0.158	0.021	4.25	2.47	0.068	0.007	0.010	0.141	0
Grape juice—frozen concentrate	51	0.184	0.088	12.60	3.68	0.102	0.015	0.026	0.492	0
Grape juice—prep frozen	15	0.053	0.026	3.62	1.13	0.029	0.004	0.007	0.142	0
Kiwi fruit	17	0.280	0.127	4.22	7.46	0.112	0.007	0.015	0.962	0
Lemon—fresh wo/peel	8	0.313	0.083	2.64	7.33	0.171	0.011	0.006	0.582	0
Lemon juice—fresh	7	0.107	0.081	2.45	2.09	0.009	0.008	0.003	0.099	0
Lemon juice—bottled	6	0.114	0.081	1.84	3.02	0.036	0.012	0.003	0.085	0
Lime—fresh	9	0.199	0.055	2.99	9.30	0.169	0.008	0.006	0.228	0
Lime juice—fresh	8	0.124	0.029	2.56	2.54	0.009	0.006	0.003	0.113	0
Lime juice—bottled	6	0.115	0.115	1.84	3.46	0.069	0.009	0.001	0.099	0
Loganberries—fresh	20	0.430	0.088	3.69	8.50	0.181	0.014	0.010	1.760	0
Loganberries—frozen	16	0.430	0.089	3.68	7.33	0.181	0.014	0.010	1.760	0
Mango—fresh—slices	19	0.146	0.077	4.83	2.92	0.360	0.016	0.016	0.997	0
Mango—fresh—whole	19	0.145	0.078	4.82	2.88	0.356	0.016	0.016	1.010	0
Cantaloupe—cubes	10	0.250	0.079	2.37	3.19	0.060	0.006	0.006	0.284	0
Casaba melon—cubes	8	0.255	0.028	1.75	1.50	0.113	0.017	0.006	0.284	0
Honeydew melon—cubes	10	0.128	0.028	2.60	1.67	0.020	0.022	0.005	0.307	0
Melon balls—mixed—frozen	9	0.239	0.023	2.25	2.79	0.008	0.005	0.006	0.295	0
Mixed fruit—dried	69	0.697	0.139	18.20	10.60	0.767	0.012	0.045	1.220	0
Mixed fruit—frozen—thawed	28	0.397	0.052	6.87	2.04	0.079	0.005	0.010	0.386	0
Nectarine	14	0.267	0.129	3.34	1.25	0.044	0.005	0.012	0.554	0
Orange	13	0.266	0.035	3.33	11.30	0.029	0.025	0.011	0.680	0
Orange sections—fresh	13	0.266	0.035	3.34	11.30	0.029	0.025	0.011	0.680	0
Mandarin oranges—canned	17	0.113	0.011	4.61	2.03	0.101	0.015	0.012	0.478	0
Orange juice—fresh	13	0.199	0.057	2.95	3.09	0.057	0.025	0.008	0.113	0
Oranged juice—frozen concentrate	45	0.679	0.059	10.80	9.05	0.099	0.079	0.018	0.313	0
Orange juice—frozen	13	0.191	0.016	3.05	2.50	0.031	0.023	0.005	0.057	0
Papaya—whole fresh	11	0.173	0.040	2.78	6.71	0.028	0.008	0.009	0.482	0
Papaya—slices fresh	12	0.174	0.040	2.77	6.68	0.060	0.008	0.009	0.482	0
Papaya nectar—canned	16	0.049	0.043	4.12	2.72	0.098	0.002	0.001	0.170	0
Peaches—fresh	12	0.199	0.026	3.14	1.30	0.031	0.005	0.012	0.489	0

FRUITS — continued

	kcal	Protein (g)	Fat (g)	CHO (g)	Ca (mg)	Fe (mg)	B₁ (mg)	B₂ (mg)	Fiber (g)	Cholesterol (mg)
Peach slices — frozen/thawed	27	0.177	0.037	6.80	0.91	0.105	0.004	0.010	0.467	0
Peach halves — heavy syrup	21	0.130	0.028	5.67	1.05	0.077	0.003	0.007	0.315	0
Peach halves — light syrup	15	0.126	0.010	4.13	1.05	0.101	0.002	0.007	0.402	0
Peach halves — dried	68	0.020	0.216	17.40	8.07	1.150	0	0.060	2.330	0
Peach nectar — canned	15	0.076	0.006	3.95	1.48	0.054	0	0.004	0.170	0
Pears — Bartlett	17	0.111	0.113	4.29	3.24	0.070	0.006	0.011	0.779	0
Pear halves — heavy syrup	21	0.057	0.036	5.42	1.44	0.061	0.004	0.006	0.395	0
Pear halves — light syrup	16	0.054	0.007	4.30	1.44	0.082	0.003	0.005	0.395	0
Pear nectar — canned	17	0.030	0.003	4.47	1.25	0.073	0	0.004	0.204	0
Pineapple slices — heavy syrup	22	0.098	0.034	5.72	3.91	0.108	0.025	0.007	0.270	0
Pineapple slices — light syrup	15	0.103	0.034	3.81	3.91	0.110	0.026	0.007	0.268	0
Pineapple — frozen sweetened	24	0.113	0.029	6.29	2.55	0.113	0.028	0.009	0.496	0
Pineapple juice — frozen concentrate	51	0.369	0.029	12.60	11.00	0.255	0.065	0.017	0.255	0
Plums	16	0.223	0.176	3.69	1.29	0.030	0.012	0.027	0.550	0
Plums — canned — heavy syrup	25	0.102	0.028	6.59	2.53	0.238	0.005	0.010	0.434	0
Plums — canned — light syrup	18	0.105	0.029	4.61	2.70	0.243	0.005	0.011	0.444	0
Prunes — dried	68	0.739	0.145	17.80	14.50	0.702	0.023	0.046	2.700	0
Prune juice — bottled	20	0.172	0.009	4.95	3.43	0.334	0.005	0.020	0.310	0
Raisins — seedless	85	0.914	0.130	22.50	13.90	0.594	0.044	0.025	1.670	0
Raspberries — fresh	14	0.256	0.157	3.27	6.22	0.162	0.009	0.026	1.770	0
Raspberries — canned w/liquid	26	0.235	0.034	6.62	2.99	0.120	0.006	0.009	1.200	0
Raspberries — frozen	29	0.197	0.044	7.42	4.30	0.184	0.005	0.013	1.300	0
Rhubarb — raw — diced	6	0.253	0.056	1.29	39.50	0.062	0.006	0.009	0.737	0
Rhubarb — cooked w/sugar	33	0.111	0.013	8.85	41.10	0.060	0.005	0.006	0.624	0
Strawberries — fresh	9	0.173	0.105	2.00	4.00	0.108	0.006	0.019	0.736	0
Strawberries — frozen	10	0.120	0.030	2.59	4.38	0.213	0.006	0.010	0.736	0
Tangerine — fresh	13	0.179	0.054	3.17	4.05	0.028	0.030	0.006	0.574	0
Tangerines — canned — light syrup	17	0.127	0.028	4.61	2.03	0.105	0.015	0.012	0.453	0
Watermelon	9	0.175	0.121	2.04	2.24	0.048	0.023	0.006	0.114	0

MEATS

	kcal	Protein (g)	Fat (g)	CHO (g)	Ca (mg)	Fe (mg)	B₁ (mg)	B₂ (mg)	Fiber (g)	Cholesterol (mg)
Beef chuck — pot roasted	108	7.20	8.64	0	3.67	0.840	0.020	0.065	0	29.0
Beef chuck — pot roasted lean	77	8.80	4.34	0	3.67	1.040	0.024	0.080	0	30.0
Beef round — pot roasted lean & fat	74	8.44	4.20	0	1.67	0.920	0.020	0.069	0	27.0
Beef round — pot roasted lean	63	8.97	2.74	0	1.33	0.980	0.021	0.074	0	27.0
Ground beef — lean	77	7.00	5.34	0	3.00	0.600	0.013	0.060	0	24.7
Ground beef — regular	82	6.67	5.94	0	3.00	0.700	0.010	0.053	0	25.3
Sirloin steak — lean	57	8.10	2.53	0	2.33	0.700	0.026	0.056	0	21.7
T-bone steak — lean & fat	92	6.80	6.97	0	2.67	0.720	0.026	0.059	0	23.7
Beef lunchmeat — thin-sliced	50	7.96	1.09	1.62	2.99	0.759	0.023	0.054	0	12.0
Beef lunchmeat — loaf/roll	87	4.06	7.42	0.82	2.99	0.659	0.030	0.062	0	18.0
Beef rib — oven roasted — lean	68	7.70	3.90	0	3.34	0.740	0.023	0.060	0	22.7
Beef round — oven roasted — lean	54	8.14	2.12	0	1.67	0.834	0.028	0.076	0	23.0
Beef rump roast — lean only	51	8.40	1.89	0	1.13	0.567	0.026	0.049	0	19.6
Beef brains — pan-fried	56	3.57	4.50	0	2.67	0.630	0.037	0.074	0	566.0
Beef heart	47	8.17	1.59	0.12	1.67	2.130	0.040	0.436	0	54.7

MEATS—continued

	kcal	Protein (g)	Fat (g)	CHO (g)	Ca (mg)	Fe (mg)	B₁ (mg)	B₂ (mg)	Fiber (g)	Cholesterol (mg)
Beef kidney	41	7.23	0.97	0.27	5.00	2.070	0.054	1.150	0	110.0
Beef liver—fried	61	7.57	2.27	2.23	3.00	1.780	0.060	1.170	0	137.0
Beef tongue—cooked	80	6.27	5.87	0.09	2.00	0.960	0.009	0.009	0	30.4
Beef tripe—raw	28	4.14	1.12	0	36.00	0.553	0.002	0.047	0	26.9
Beef tripe—pickled	17	3.29	0.40	0	25.00	0.389	0	0.028	0	15.0
Corned beef—canned	71	7.67	4.24	0	5.67	0.590	0.006	0.042	0	24.3
Corned beef hash—canned	49	2.35	1.29	2.82	3.74	0.567	0.016	0.052	0.153	17.0
Beef—dried/cured	47	8.24	1.10	0.44	2.00	1.280	0.050	0.230	0	45.9
Beef & vegetable stew	26	1.85	1.27	1.74	3.36	0.336	0.017	0.020	0.393	8.2
Beef stew—canned	22	1.64	0.88	2.06	2.66	0.368	0.008	0.014	0.150	1.7
Burrito—beef & bean	63	3.40	2.84	6.48	26.70	0.437	0.042	0.047	0.810	8.4
Tostada w/beans & beef	49	2.72	3.06	2.98	27.50	0.319	0.012	0.036	0.583	9.1
Beef+macaroni+tomato	24	1.25	0.73	3.16	3.81	0.300	0.024	0.021	0.291	2.8
Beef enchilada	69	3.09	3.26	3.69	60.20	0.418	0.017	0.039	0.465	8.9
Frankfurter—beef	92	3.20	8.36	0.68	3.48	0.378	0.014	0.029	0	13.4
Frankfurter—beef & pork	91	3.20	8.26	0.73	2.98	0.328	0.056	0.034	0	14.4
Beef pot pie—frozen	52	1.99	2.73	4.77	2.42	0.436	0.022	0.018	0.109	5.0
Beef pie—recipe	70	2.84	4.05	5.27	3.92	0.513	0.039	0.039	0.155	5.7
Beef taco	75	4.94	4.80	3.67	30.90	0.469	0.010	0.049	0.407	16.2
Chicken meat—all-fried	62	8.67	2.59	0.48	4.86	0.383	0.024	0.056	0.002	26.5
Chicken meat—all-roasted	54	8.20	2.10	0	4.25	0.342	0.020	0.050	0	25.3
Chicken meat—all-stewed	50	7.74	1.90	0	4.05	0.330	0.014	0.046	0	23.5
Boned chicken w/broth	47	6.17	2.26	0	3.99	0.439	0.004	0.037	0	17.6
Chicken—dark meat—fried	68	8.22	3.30	0.74	5.06	0.423	0.026	0.070	0.002	27.3
Chicken—dark meat—roasted	58	7.76	2.75	0	4.25	0.377	0.020	0.064	0	26.3
Chicken—dark meat—stewed	55	7.37	2.55	0	4.05	0.385	0.016	0.057	0	24.9
Chicken—light meat—fried	54	9.31	1.57	0.12	4.45	0.322	0.020	0.036	0	25.3
Chicken—light meat—roasted	49	8.77	1.28	0	4.25	0.302	0.018	0.033	0	23.9
Chicken—light meat—stewed	45	8.18	1.17	0	3.64	0.265	0.012	0.033	0	21.7
Chicken breast—no skin	47	8.80	0.99	0	4.29	0.295	0.020	0.032	0	24.0
Chicken breast meat—stewed	43	8.20	0.86	0	3.58	0.250	0.012	0.034	0	21.8
Chicken drumstick—batter fried	76	6.22	4.45	2.36	4.73	0.382	0.032	0.061	0.008	24.4
Chicken drumstick—roasted	61	7.69	3.16	0	3.27	0.376	0.020	0.061	0	26.2
Chicken wing—batter-fried	92	5.64	6.19	3.10	5.79	0.365	0.030	0.043	0.012	22.6
Chicken wing—flour-fried	91	7.40	6.28	0.67	4.43	0.354	0.017	0.039	0.009	23.0
Chicken wing—roasted	83	7.48	5.53	0	4.17	0.359	0.012	0.037	0	24.2
Chicken gizzards—simmered	44	7.73	1.04	0.32	2.58	1.180	0.008	0.070	0	54.9
Chicken hearts—simmered	52	7.49	2.23	0.03	5.27	2.560	0.017	0.206	0	68.7
Chicken livers—simmered	44	6.90	1.55	0.25	4.05	2.400	0.043	0.496	0	179.0
Chicken roll—light meat	45	5.52	2.08	0..69	11.90	0.274	0.018	0.037	0	13.9
Chicken frankfurter	73	3.67	5.52	1.93	27.00	0.567	0.019	0.033	0	28.4
Chicken a la king	54	3.12	3.93	1.93	14.70	0.289	0.012	0.049	0.154	25.6
Chicken+noodles	43	2.60	2.13	3.07	3.07	0.278	0.006	0.020	0.142	12.2
Chicken chow mein	29	2.60	1.25	1.13	6.58	0.284	0.009	0.026	0.466	8.5
Chicken curry	26	2.21	1.53	0.64	2.52	0.170	0.009	0.019	0.033	5.3
Chicken frankfurter	72	3.67	5.52	1.93	27.00	0.567	0.020	0.033	0	28.4
Chicken pot pie—frozen	53	1.84	2.84	5.08	3.70	0.382	0.020	0.020	0.210	4.9
Chicken roll—light meat	45	5.52	2.08	0.69	11.90	0.274	0.018	0.037	0	13.9
Chicken salad w/celery	97	3.82	8.90	0.47	5.92	0.239	0.012	0.028	0.109	17.3
Chicken patty sandwich	79	4.48	4.06	6.10	7.95	0.338	0.052	0.047	0.244	12.3
Chicken broth—from dry	2	0.16	0.13	0.17	1.74	0.009	0	0.004	0.001	0.1
Chicken broth—from cube	2	0.11	0.04	0.18	0.04	0.014	0.001	0.003	0	0.1
Chicken noodle soup	9	0.48	0.29	1.10	2.00	0.092	0.006	0.007	0.085	0.8
Tostada w/beans/chicken	45	3.50	2.06	3.38	29.30	0.305	0.013	0.034	0.668	9.6

MEATS—continued

	kcal	Protein (g)	Fat (g)	CHO (g)	Ca (mg)	Fe (mg)	B₁ (mg)	B₂ (mg)	Fiber (g)	Cholesterol (mg)
Chicken taco	63	5.60	3.03	3.67	31.60	0.237	0.014	0.043	0.407	16.5
Chicken enchilada	64	3.38	2.48	3.69	60.50	0.359	0.018	0.036	0.465	9.1
Turkey dark meat—roasted	53	8.10	2.05	0	9.11	0.662	0.018	0.070	0	24.0
Turkey white meat—roasted	44	8.48	0.91	0	5.47	0.380	0.017	0.037	0	19.6
Turkey breast—barbecued	40	6.39	1.40	0	2.00	0.120	0.010	0.030	0	16.0
Turkey gizzards	46	8.34	1.10	0.17	4.32	1.540	0.009	0.093	0	65.6
Turkey hearts	50	7.58	1.73	0.58	3.72	1.950	0.019	0.250	0	64.0
Turkey livers	48	6.80	1.69	0.97	3.02	2.210	0.015	0.404	0	177.0
Turkey loaf	31	6.38	0.45	0	2.00	0.113	0.011	0.030	0	11.6
Turkey roll	41	5.27	2.03	0.15	11.40	0.358	0.025	0.064	0	11.9
Turkey bologna	56	3.86	4.27	0.27	23.40	0.432	0.015	0.047	0	28.0
Turkey frankfurter	64	4.05	5.22	0.42	36.50	0.485	0.023	0.050	0	24.6
Turkey ham	36	5.37	1.49	0.42	2.49	0.776	0.020	0.075	0	15.9
Turkey pastrami	37	5.22	1.75	0.43	2.49	0.403	0.022	0.075	0	14.9
Turkey salami	55	4.62	3.89	0.15	5.47	0.463	0.029	0.075	0	22.9
Turkey pot pie—frozen	51	1.80	2.75	4.65	7.79	0.256	0.020	0.020	0.110	2.4

EGGS

	kcal	Protein (g)	Fat (g)	CHO (g)	Ca (mg)	Fe (mg)	B₁ (mg)	B₂ (mg)	Fiber (g)	Cholesterol (mg)
Egg white, cooked	13	2.68	0	0.33	3.20	0.008	0.002	0.072	0	0
Egg yolk, cooked	108	4.76	8.73	0.07	44.40	1.620	0.044	0.121	0	355.0
Egg, fried in butter	56	3.54	3.45	0.38	15.70	0.536	0.018	0.148	0	129.0
Egg, hard cooked	40	3.52	2.79	0.34	13.80	0.474	0.014	0.132	0	113.0
Egg, poached	40	3.50	2.79	0.34	13.80	0.474	0.014	0.132	0	113.0
Egg, scrambled milk + butter	40	2.88	2.57	0.61	23.90	0.412	0.013	0.106	0	93.9
Egg raw—large	40	3.52	2.79	0.34	13.80	0.474	0.017	0.139	0	113.0
Egg white—raw	13	2.68	0	0.33	3.20	0.008	0.002	0.075	0	0
Egg yolk, raw	108	4.76	8.73	0.07	44.40	1.620	0.051	0.126	0	355.0
Egg substitute, frozen	45	3.20	3.15	0.91	20.80	0.562	0.034	0.110	0	0.5
Egg substitute, powder	125	15.60	3.66	6.12	90.70	0.879	0.062	0.493	0	162.0

DAIRY PRODUCTS

	kcal	Protein (g)	Fat (g)	CHO (g)	Ca (mg)	Fe (mg)	B₁ (mg)	B₂ (mg)	Fiber (g)	Cholesterol (mg)
Milk—1% lowfat	12	0.93	0.30	1.36	34.9	0.014	0.011	0.047	0	1.16
Milk—2% lowfat	14	0.94	0.56	1.36	34.5	0.012	0.011	0.047	0	2.56
Milk—skim	10	0.97	0.05	1.38	34.9	0.012	0.010	0.040	0	0.46
Milk—whole	17	0.93	0.95	1.32	33.8	0.014	0.010	0.046	0	3.83
Buttermilk	12	0.94	0.25	1.35	33.0	0.014	0.010	0.044	0	1.04
Milk—instant nonfat dry	102	9.96	0.21	14.80	349.0	0.088	0.117	0.496	0	5.00
Canned skim milk—evaporated	22	2.11	0.06	3.22	82.0	0.078	0.013	0.088	0	1.11
Canned whole milk—evaporated	38	1.91	2.20	2.81	73.9	0.054	0.014	0.090	0	8.33
Carob flavor mix—powder	106	0.47	0.05	26.50	0	1.300	0.002	0	4.020	0
Chocolate milk—1%	18	0.92	0.28	2.96	32.5	0.068	0.010	0.047	0.425	0.79
Chocolate milk—2%	20	0.91	0.57	2.95	32.2	0.068	0.010	0.046	0.425	1.93
Chocolate milk—whole	24	0.90	0.96	2.94	31.8	0.068	0.010	0.046	0.425	3.52
Hot cocoa—with whole milk	25	1.03	1.03	2.93	33.8	0.088	0.012	0.049	0.340	3.74
Instant breakfast w/2% milk	25	1.52	0.47	3.50	31.0	0.807	0.040	0.048	0	1.82
Instant breakfast w/1% milk	23	1.51	0.25	3.50	31.3	0.807	0.040	0.048	0	1.00

DAIRY PRODUCTS—continued

	kcal	Protein (g)	Fat (g)	CHO (g)	Ca (mg)	Fe (mg)	B₁ (mg)	B₂ (mg)	Fiber (g)	Cholesterol (mg)
Instant breakfast w/skim milk	22	1.55	0.04	3.50	31.4	0.804	0.039	0.041	0	0.40
Instant breakfast w/whole milk	28	1.51	0.82	3.47	30.4	0.807	0.040	0.047	0	3.33
Egg nog	38	1.08	2.12	3.84	36.8	0.057	0.010	0.054	0	16.60
Kefir	20	1.13	0.55	1.07	42.6	0.060	0.055	0.054	0	1.22
Malt powder—chocolate flavored	107	1.49	1.08	24.80	17.6	0.648	0.049	0.057	0.540	1.35
Malted milk powder	117	3.12	2.30	21.50	85.0	0.209	0.143	0.260	0.405	5.40
Malted milk drink—chocolate	25	1.00	0.95	3.19	32.5	0.064	0.014	0.047	0.043	3.64
Chocolate milkshake	36	0.96	1.05	5.80	32.0	0.088	0.016	0.069	0.035	3.70
Strawberry milkshake	32	0.95	0.80	5.35	32.0	0.030	0.013	0.055	0.024	3.10
Vanilla milkshake	32	0.98	0.84	5.09	34.5	0.026	0.013	0.052	0.019	3.20
Ovaltine powder—chocolate flavored	102	2.00	0.85	23.70	134.0	6.190	0.719	0.772	0.013	0
Ovaltine powder—malt flavored	104	2.53	0.24	23.60	106.0	5.820	0.772	1.010	0.040	0
Ovaltine drink—chocolate flavored	24	1.02	0.94	3.12	41.9	0.510	0.067	0.104	0.001	3.53
Ovaltine drink—malt flavored	24	1.06	0.89	3.10	39.7	0.480	0.072	0.124	0.003	3.53
Milk—goat	20	1.00	1.17	1.27	37.9	0.014	0.014	0.039	0	3.25
Milk—sheep	31	1.70	1.99	1.52	54.8	0.028	0.018	0.100	0	0
Milk—soybean	9	0.78	0.54	0.51	1.2	0.163	0.046	0.020	0	0
Ice cream—regular-vanilla	57	1.02	3.05	6.76	37.5	0.026	0.011	0.070	0	12.60
Ice cream—rich-vanilla	67	0.79	4.54	6.13	28.9	0.019	0.008	0.054	0	16.90
Ice cream—soft-serve	62	1.15	3.69	6.28	38.7	0.070	0.013	0.073	0	25.00
Creamsicle ice cream bar	44	0.52	1.33	7.56	19.8	0	0.009	0.034	0	0
Drumstick ice cream bar	88	1.23	4.68	10.20	31.7	0.047	0.009	0.043	0	0
Fudgesicle ice cream bar	35	1.48	0.08	7.22	50.0	0.039	0.012	0.070	0	0
Ice milk	40	1.12	1.22	6.28	38.0	0.039	0.016	0.075	0	3.90
Ice milk—soft serve—3% fat	36	1.30	0.75	6.22	44.4	0.045	0.019	0.088	0	2.10
Yogurt—coffee-vanilla	24	1.40	0.35	3.90	48.5	0.020	0.012	0.057	0	1.42
Yogurt—lowfat with fruit	29	1.24	0.31	5.37	43.0	0.020	0.010	0.050	0	1.25
Yogurt—lowfat-plain	18	1.49	0.43	2.00	51.8	0.022	0.012	0.060	0	1.75
Yogurt—nonfat milk	16	1.62	0.05	2.17	56.5	0.025	0.014	0.066	0	0.50
Yogurt—whole milk	17	0.98	0.92	1.32	34.3	0.014	0.008	0.040	0	3.68

VEGETABLES

	kcal	Protein (g)	Fat (g)	CHO (g)	Ca (mg)	Fe (mg)	B₁ (mg)	B₂ (mg)	Fiber (g)	Cholesterol (mg)
Alfalfa sprouts	9	1.13	0.200	1.07	9.5	0.272	0.021	0.036	1.030	0
Artichoke hearts—marinated	28	0.680	0.250	2.18	6.5	0.270	0.010	0.029	1.780	0
Asparagus—raw spears	6	0.865	0.063	1.05	6.4	0.193	0.032	0.035	0.395	0
Asparagus—canned spears	5	0.606	0.184	0.70	3.9	0.177	0.017	0.025	0.454	0
Bamboo shoots—sliced—raw	8	0.738	0.088	1.47	3.8	0.143	0.043	0.020	0.738	0
Bamboo shoots—sliced, canned	5	0.489	0.113	0.91	2.2	0.090	0.007	0.007	0.706	0
Bean sprouts—fresh raw	9	0.861	0.050	1.68	3.8	0.258	0.024	0.035	0.736	0
Bean sprouts—boiled	6	0.576	0.025	1.19	3.4	0.185	0.014	0.029	0.572	0
Bean sprouts—stir fried	14	1.220	0.059	3.00	3.7	0.549	0.040	0.050	0.777	0
Black beans—cooked	37	2.500	0.152	6.72	7.8	0.593	0.069	0.017	2.540	0
Green beans—raw uncooked	9	0.515	0.034	2.02	10.6	0.363	0.024	0.030	0.644	0
Green beans—fresh—cooked	10	0.535	0.082	2.24	13.2	0.363	0.021	0.027	0.737	0

VEGETABLES—continued

	kcal	Protein (g)	Fat (g)	CHO (g)	Ca (mg)	Fe (mg)	B₁ (mg)	B₂ (mg)	Fiber (g)	Cholesterol (mg)
Green beans—frozen—cooked	8	0.386	0.038	1.73	12.8	0.233	0.014	0.021	0.880	0
Green beans—canned/drained	6	0.326	0.028	1.28	7.6	0.256	0.004	0.016	0.378	0
Red kidney beans—dry	94	6.690	0.234	16.90	40.5	2.330	0.150	0.062	6.160	0
Lima beans—dry large	96	6.080	0.194	18.00	22.9	2.130	0.144	0.057	8.600	0
Lima beans—fresh—cooked	35	1.930	0.090	6.70	9.0	0.695	0.040	0.027	2.670	0
Lima beans—dry small	98	5.790	0.397	18.20	18.4	2.270	0.136	0.048	8.500	0
Lima beans—canned/drained	27	1.530	0.100	5.20	8.0	0.487	0.010	0.013	2.420	0
Beans/w/franks—canned	40	1.900	1.860	4.37	13.6	0.490	0.016	0.016	1.950	1.7
Pork & beans—canned	32	1.500	0.413	5.95	17.4	0.470	0.013	0.017	1.560	1.9
Navy beans—dry, cooked	40	2.460	0.162	7.77	19.9	0.703	0.057	0.017	2.490	0
Pinto beans—dry, cooked	39	2.320	0.148	7.28	13.6	0.741	0.053	0.026	3.230	0
Refried beans—canned	30	1.770	0.303	5.24	13.2	0.500	0.014	0.016	2.470	0
Soybeans—dry	118	10.400	5.650	8.55	78.5	4.450	0.248	0.247	1.570	0
White beans—dry	95	5.990	0.334	17.70	36.0	2.190	0.210	0.059	0.766	0
White beans—dry, cooked	40	2.550	0.182	7.32	20.7	0.808	0.067	0.017	2.230	0
Yellow wax beans—raw	9	0.515	0.034	2.02	10.6	0.294	0.024	0.030	0.644	0
Yellow wax beans—raw	10	0.535	0.082	2.24	13.2	0.363	0.021	0.027	0.726	0
Yellow wax beans—frozen	8	0.386	0.038	1.73	12.8	0.233	0.014	0.021	0.880	0
Beets—cooked	9	0.300	0.014	1.90	3.1	0.176	0.009	0.004	0.539	0
Beets—pickled slices	18	0.227	0.028	4.63	3.1	0.116	0.006	0.014	0.587	0
Broccoli—raw chopped	8	0.844	0.097	1.49	13.5	0.251	0.019	0.034	0.934	0
Broccoli—raw spears	8	0.845	0.098	1.49	13.5	0.250	0.018	0.034	0.935	0
Broccoli—frozen cooked spears	8	0.879	0.034	1.51	14.5	0.173	0.015	0.023	0.826	0
Brussels sprouts—raw	12	0.960	0.084	2.54	11.6	0.396	0.039	0.026	1.260	0
Brussels sprouts—cooked	11	1.090	0.145	2.45	10.2	0.342	0.030	0.023	1.220	0
Brussels sprouts—frozen cooked	12	1.030	0.112	2.36	7.0	0.210	0.029	0.032	1.230	0
Cabbage—raw, shredded	7	0.340	0.049	1.52	13.0	0.162	0.015	0.009	0.680	0
Cabbage—cooked	6	0.272	0.070	1.35	9.5	0.110	0.016	0.016	0.661	0
Bok choy—raw, shredded	4	0.425	0.057	0.62	30.0	0.227	0.011	0.020	0.486	0
Bok choy—cooked	3	0.442	0.045	0.51	26.3	0.295	0.009	0.018	0.454	0
Red cabbage—raw	8	0.393	0.073	1.74	14.6	0.142	0.018	0.009	0.648	0
Red cabbage—cooked	6	0.299	0.057	1.32	10.6	0.102	0.010	0.006	1.567	0
Carrot—whole, raw	12	0.291	0.055	2.87	7.5	0.142	0.028	0.017	1.906	0
Carrot—grated, raw	12	0.289	0.052	2.88	7.7	0.142	0.027	0.016	0.907	0
Carrots—sliced, cooked	13	0.309	0.050	2.97	8.7	0.176	0.010	0.016	0.992	0
Carrots—frozen, cooked	10	0.338	0.031	2.34	8.2	0.136	0.008	0.010	1.050	0
Carrots—canned, drained	7	0.183	0.054	1.57	7.4	0.181	0.005	0.009	0.435	0
Carrot juice	11	0.267	0.041	2.63	6.7	0.130	0.026	0.015	0.385	0
Cauliflower—raw	7	0.561	0.051	1.39	7.9	0.164	0.022	0.016	0.720	0
Cauliflower—cooked	7	0.530	0.050	1.31	7.8	0.119	0.018	0.018	0.622	0
Cauliflower—frozen, cooked	5	0.457	0.061	1.06	4.9	0.116	0.010	0.015	0.535	0
Celery—raw—chopped	5	0.189	0.033	1.03	10.4	0.137	0.009	0.009	0.472	0
Swiss chard—raw	5	0.510	0.057	1.06	14.5	0.510	0.011	0.025	0.512	0
Swiss chard—cooked	6	0.533	0.023	1.17	16.5	0.642	0.010	0.024	0.616	0
Collards—fresh	5	0.445	0.062	1.07	33.0	0.299	0.009	0.018	0.590	0
Collards—fresh, cooked	4	0.313	0.043	0.75	22.0	0.116	0.005	0.012	0.798	0
Collards—frozen, cooked	10	0.840	0.115	2.02	59.5	0.317	0.013	0.033	0.794	0
Corn—kernels raw	24	0.913	0.335	5.38	0.6	0.147	0.057	0.017	1.220	0
Corn on the cob—cooked	31	0.943	0.363	7.14	0.6	0.173	0.061	0.020	1.190	0
Corn—cooked from frozen	23	0.857	0.020	5.80	0.6	0.085	0.020	0.020	1.190	0
Corn—canned, drained	23	0.743	0.283	5.26	1.4	0.142	0.009	0.014	0.398	0
Corn—canned cream style	21	0.494	0.119	5.14	0.9	0.108	0.007	0.015	0.354	0
Cucumber slices w/peel	4	0.153	0.037	0.82	4.0	0.079	0.009	0.006	0.329	0
Eggplant—cooked	8	0.236	0.066	1.88	1.7	0.099	0.022	0.006	1.060	0
Escarole/curly endive—chopped	5	0.354	0.057	0.95	14.7	0.235	0.023	0.022	0.369	0
Garbanzo/chickpeas—dry	103	5.470	1.720	17.20	29.9	1.770	0.135	0.060	5.390	0
Garbanzo/chickpeas—cooked	47	2.500	0.735	7.78	13.8	0.819	0.033	0.018	1.920	0
Jerusalem artichoke—raw	22	0.567	0.004	4.95	4.0	0.964	0.057	0.017	0.369	0
Kale—fresh, chopped	14	0.935	0.198	2.84	38.0	0.482	0.031	0.037	1.650	0

VEGETABLES — continued

	kcal	Protein (g)	Fat (g)	CHO (g)	Ca (mg)	Fe (mg)	B₁ (mg)	B₂ (mg)	Fiber (g)	Cholesterol (mg)
Kohlrabi — raw slices	8	0.482	0.028	1.76	6.9	0.113	0.014	0.006	0.405	0
Kohlrabi — cooked	8	0.510	0.030	1.96	7.0	0.113	0.011	0.006	0.395	0
Leeks — chopped raw	17	0.425	0.085	4.00	16.7	0.594	0.017	0.008	0.668	0
Leeks — cooked, chopped	9	0.230	0.057	2.16	8.5	0.310	0.007	0.006	0.927	0
Lentils — dry	96	7.960	0.273	16.20	14.6	2.550	0.135	0.069	3.400	0
Lentils — cooked from dry	33	2.560	0.106	5.73	5.3	0.944	0.048	0.020	1.430	0
Lentils — sprouted, raw	30	2.540	0.156	6.30	7.0	0.909	0.065	0.036	1.150	0
Lettuce — butterhead	4	0.367	0.062	0.66	9.5	0.085	0.017	0.017	0.397	0
Lettuce — iceberg	4	0.286	0.054	0.59	5.4	0.142	0.013	0.009	0.347	0
Lettuce — Romaine	5	0.459	0.057	0.67	10.2	0.312	0.028	0.028	0.482	0
Mushrooms — raw sliced	7	0.593	0.119	1.32	1.4	0.352	0.029	0.127	0.508	0
Mushrooms — cooked	8	0.614	0.134	1.46	1.7	0.494	0.020	0.085	0.625	0
Mushrooms — canned, drained	7	0.530	0.082	1.40	3.1	0.224	0.017	0.063	0.596	0
Mustard greens — fresh	7	0.764	0.057	1.39	29.4	0.414	0.023	0.031	0.764	0
Mustard greens — cooked	4	0.640	0.068	0.60	21.0	0.316	0.012	0.018	0.587	0
Okra pods — cooked	9	0.530	0.048	2.04	17.9	0.128	0.037	0.016	0.624	0
Okra slices — cooked	11	0.589	0.085	2.32	27.1	0.190	0.028	0.035	0.709	0
Onions — chopped, raw	10	0.335	0.074	2.07	7.1	0.105	0.017	0.003	0.454	0
Onion slices — raw	10	0.335	0.074	2.08	7.2	0.105	0.017	0.003	0.454	0
Onion — dehydrated flakes	91	2.530	0.122	23.70	72.9	0.446	0.022	0.018	2.510	0
Onion rings — frozen, heated	115	1.520	7.570	10.80	8.8	0.482	0.079	0.040	0.240	0
Parsley — freeze dried	81	8.910	1.420	11.90	40.5	15.200	0.304	0.648	22.000	0
Parsley — fresh chopped	9	0.624	0.085	1.96	36.9	1.760	0.023	0.031	1.550	0
Parsnips — sliced raw	21	0.341	0.085	5.09	10.0	0.167	0.026	0.014	1.280	0
Fresh peas — uncooked	23	1.530	0.113	4.10	7.0	0.416	0.075	0.037	1.380	0
Peas — cooked	24	1.520	0.060	4.43	7.8	0.438	0.073	0.042	1.360	0
Peas — frozen, cooked	22	1.460	0.078	4.04	6.7	0.443	0.080	0.050	1.280	0
Peas — edible pods — fresh	12	0.794	0.057	2.15	12.1	0.589	0.043	0.023	0.794	0
Split peas — dry	97	6.970	0.328	17.10	15.5	1.250	0.206	0.061	4.030	0
Peas + carrots — frozen, cooked	14	0.875	0.120	2.87	6.4	0.266	0.064	0.020	1.170	0
Green chili pepper — raw	11	0.557	0.057	2.68	5.0	0.340	0.026	0.026	0.504	0
Red chili peppers — raw/ chopped	11	0.567	0.057	2.68	4.9	0.340	0.026	0.026	0.454	0
Jalapeno peppers — canned chopped	7	0.225	0.170	1.39	7.5	0.792	0.008	0.014	0.850	0
Baked potato with skin	31	0.653	0.028	7.16	2.8	0.386	0.030	0.009	0.660	0
Baked potato — flesh only	26	0.556	0.029	6.10	1.5	0.100	0.030	0.006	0.436	0
Potato skin — oven baked	56	1.220	0.029	13.20	9.8	1.080	0.035	0.034	1.130	0
Potato + peel — microwaved	30	0.692	0.028	6.83	3.1	0.350	0.034	0.009	0.660	0
Peeled potato — boiled	24	0.485	0.029	5.67	2.1	0.088	0.028	0.005	0.426	0
French fries — oven heated	63	0.980	2.480	9.64	2.3	0.380	0.035	0.009	0.567	0
French fries — frozen — vegetable oil	90	1.140	4.690	11.20	5.7	0.215	0.050	0.008	0.567	0
Cottage-fried potatoes	62	0.975	2.320	9.64	2.8	0.425	0.034	0.009	0.567	0
Hash-brown potatoes	30	0.343	1.980	3.02	1.1	0.115	0.010	0.003	0.567	0
Mashed potatoes prep/ milk	22	0.548	0.166	4.98	7.4	0.077	0.025	0.011	0.405	0.5
Mashed potatoes — milk & margarine	30	0.533	1.200	4.74	7.3	0.074	0.024	0.014	0.405	0.5
Potato pancakes	88	1.730	4.700	9.85	7.8	0.451	0.039	0.035	0.563	34.7
Potatoes au gratin mix	26	0.653	1.170	3.65	23.5	0.090	0.006	0.023	0.487	1.4
Scalloped potatoes — recipe	24	0.813	1.040	3.05	16.2	0.163	0.020	0.026	0.289	3.4
Potato chips	148	1.590	10.000	14.70	7.0	0.339	0.040	0.006	1.360	0
Potato flour	100	2.260	0.226	22.60	9.3	4.880	0.119	0.040	0.317	0
Pumpkin — canned	10	0.311	0.080	2.28	7.4	0.394	0.007	0.015	0.519	0
Red radishes	4	0.170	0.151	1.01	5.7	0.082	0.001	0.013	0.624	0
Rutabaga — cooked cubes	10	0.314	0.053	2.19	12.0	0.133	0.020	0.010	0.434	0
Sauerkraut — canned liquid	5	0.258	0.040	1.21	8.7	0.417	0.006	0.006	0.529	0
Soybeans — mature, raw	37	3.700	1.900	3.17	19.4	0.599	0.096	0.033	0.656	0
Spinach — cooked from fresh	7	0.843	0.074	1.06	38.4	1.010	0.027	0.067	0.702	0

VEGETABLES — continued

	kcal	Protein (g)	Fat (g)	CHO (g)	Ca (mg)	Fe (mg)	B$_1$ (mg)	B$_2$ (mg)	Fiber (g)	Cholesterol (mg)
Summer squash — raw slices	6	0.334	0.061	1.23	5.7	0.130	0.018	0.010	0.425	0
Zucchini squash — cooked	5	0.181	0.014	1.11	3.6	0.099	0.012	0.012	0.567	0
Acorn squash — boiled/ mashed	10	0.190	0.023	2.49	7.5	0.159	0.028	0.002	0.680	0
Butternut squash/baked — cube	12	0.256	0.026	2.97	11.6	0.170	0.020	0.005	0.794	0
Spaghetti squash — baked/ boiled	8	0.187	0.073	1.83	6.0	0.095	0.010	0.006	0.794	0
Winter squash — boiled	10	0.319	0.050	2.40	5.2	0.136	0.018	0.006	0.794	0
Sweet potato — baked in skin	29	0.487	0.032	6.89	8.0	0.129	0.020	0.036	0.850	0
Candied sweet potatoes	39	0.246	0.920	7.91	7.3	0.324	0.005	0.012	0.545	0
Tofu (soybean curd)	22	2.290	1.360	0.53	29.7	1.520	0.023	0.015	0.343	0
Tomato — fresh whole	6	0.251	0.060	1.23	2.1	0.136	0.017	0.014	0.415	0
Tomatoes — whole canned	6	0.265	0.070	1.22	7.4	0.171	0.013	0.009	0.298	0
Tomato sauce — canned	9	0.376	0.047	2.04	3.9	0.218	0.019	0.016	0.425	0
Tomato paste — canned	24	1.070	0.249	5.33	9.9	0.848	0.044	0.054	1.210	0
Tomato juice — canned	5	0.215	0.017	1.20	2.6	0.164	0.013	0.009	0.220	0
Turnip cubes — raw	8	0.255	0.028	1.76	8.5	0.085	0.011	0.009	0.587	0
Mixed vegetables — frozen, cooked	17	0.812	0.043	3.70	7.2	0.232	0.020	0.034	1.120	0
Vegetable juice cocktail	6	0.178	0.026	1.29	3.1	0.119	0.012	0.008	0.178	0
Water chestnuts — raw	30	0.398	0.027	6.77	3.2	0.170	0.040	0.057	0.869	0
Watercress — fresh	3	0.650	0.033	0.37	33.4	0.050	0.025	0.033	0.719	0
White yams — raw	34	0.438	0.049	7.90	8.7	0.153	0.032	0.009	0.822	0

SALAD BAR

	kcal	Protein (g)	Fat (g)	CHO (g)	Ca (mg)	Fe (mg)	B$_1$ (mg)	B$_2$ (mg)	Fiber (g)	Cholesterol (mg)
Alfalfa sprouts	9	1.130	0.196	1.07	9.5	0.272	0.021	0.036	1.030	0
Artichoke hearts, marinated	28	0.680	2.250	2.18	6.5	0.270	0.010	0.029	1.780	0
Asparagus	6	0.867	0.062	1.05	6.4	0.193	0.032	0.035	0.432	0
Avocado	46	0.563	4.340	2.10	3.1	0.284	0.030	0.035	2.720	0
Bacon, regular	163	8.640	14.000	0.16	3.0	0.482	0.195	0.080	0	23.9
Bean sprouts	9	0.861	0.050	1.68	3.8	0.258	0.024	0.035	0.7365	0
Beets	9	0.300	0.013	1.90	3.0	0.176	0.009	0.004	0.567	0
Beets, canned, diced	9	0.260	0.040	2.04	4.3	0.517	0.003	0.012	0.590	0
Broccoli, raw	8	0.844	0.097	1.49	13.5	0.251	0.019	0.034	0.934	0
Cabbage	6	0.340	0.049	1.52	13.0	0.162	0.015	0.009	0.680	0
Cabbage, red	8	0.393	0.073	1.74	14.6	0.142	0.018	0.009	0.648	0
Carrots, grated	12	0.289	0.052	2.88	7.7	0.142	0.027	0.016	0.907	0
Cauliflower	7	0.561	0.051	1.39	7.9	0.164	0.022	0.016	0.720	0
Celery	5	0.189	0.030	1.03	10.4	0.137	0.009	0.009	0.472	0
Chicken salad	97	3.820	8.900	0.47	5.9	0.239	0.012	0.028	0.109	17.3
Crab, cooked	24	5.050	0.561	0.14	12.9	0.104	0.012	0.044	0	18.0
Croutons, dry bread cubes	105	3.690	1.040	20.50	35.0	1.020	0.099	0.099	0.085	0
Cucumber slices	4	0.153	0.037	0.82	4.0	0.079	0.009	0.006	0.329	0
Egg, chopped	40	3.520	2.790	0.34	13.8	0.475	0.014	0.133	0	113
Escarole/curly endive	5	0.354	0.057	0.95	14.7	0.235	0.023	0.022	0.369	0
Garbanzo/chickpeas, cooked	47	2.500	0.735	7.78	13.8	0.819	0.033	0.018	1.920	0
Green pepper, sweet	7	0.244	0.130	1.50	1.7	0.357	0.024	0.014	0.454	0
Ham salad	52	5.000	3.000	0.88	2.0	0.280	0.244	0.072	0	16
Ham, minced	74	4.620	5.860	0.53	2.7	0.216	0.203	0.054	0	20.2
Leeks	17	0.425	0.085	4.00	16.7	0.594	0.017	0.008	0.688	0
Lettuce, butterhead	4	0.366	0.062	0.66	9.5	0.085	0.017	0.017	0.370	0
Lettuce, iceberg	4	0.287	0.054	0.59	5.4	0.142	0.013	0.009	0.370	0
Lettuce, loose leaf	5	0.369	0.085	0.99	19.2	0.397	0.014	0.023	0.391	0
Lettuce, Romaine	5	0.459	0.057	0.67	10.2	0.312	0.028	0.028	0.482	0
Lobster meat	28	5.800	0.168	0.36	17.2	0.111	0.020	0.019	0	20.3

SALAD BAR—continued

	kcal	Protein (g)	Fat (g)	CHO (g)	Ca (mg)	Fe (mg)	B$_1$ (mg)	B$_2$ (mg)	Fiber (g)	Cholesterol (mg)
Mushrooms, raw	7	0.593	0.119	1.32	1.4	0.352	0.029	0.127	0.508	0
Onions	16	0.335	0.074	2.07	7.1	0.105	0.017	0.003	0.454	0
Parmesan cheese, grated	129	11.800	8.500	1.06	389	0.270	0.013	0.109	0	22.0
Peas, cooked	22	1.460	0.078	4.04	6.7	0.443	0.080	0.050	1.280	0
Sesame seed kernels, dried	167	7.480	15.500	2.66	37.2	2.210	0.204	0.024	1.950	0
Shrimp, boiled	28	5.930	0.306	0	11.0	0.876	0.009	0.009	0	55.3
Spinach, fresh	6	0.810	0.099	0.99	28.0	0.770	0.022	0.054	0.947	0
Sunflower seeds, dry	162	6.460	14.000	5.32	32.9	1.920	0.650	0.070	1.970	0
Tomatoes	6	0.252	0.061	1.23	1.9	0.135	0.017	0.014	0.416	0
Tuna salad	53	4.550	2.630	2.67	4.8	0.282	0.009	0.019	0.340	3.73
Turkey meat	41	5.270	2.030	0.15	11.4	0.358	0.025	0.064	0	11.9

VARIETY

	kcal	Protein (g)	Fat (g)	CHO (g)	Ca (mg)	Fe (mg)	B$_1$ (mg)	B$_2$ (mg)	Fiber (g)	Cholesterol (mg)
Chips & crackers										
Doritos—nacho flavor	139	2.20	6.79	18.00	17.0	0.399	0.040	0.030	1.10	0
Doritos—taco flavor	140	2.60	6.59	17.60	44.9	0.699	0.080	0.090	1.10	0
Potato chips—sour cream & onion	153	2.40	9.48	14.60	21.0	0.474	0.040	0.055	1.35	1.0
Wheat cracker—thin	124	3.19	4.96	17.70	10.6	1.060	0.142	0.106	1.84	0
Whole wheat crackers	124	3.19	5.32	17.70	10.6	1.850	0.070	0.106	2.94	0
Condiments										
Catsup	30	0.52	0.11	7.17	6.2	0.228	0.026	0.020	0.45	0
Mustard	21	1.34	1.25	1.81	23.8	0.567	0.024	0.057	0.11	0
Soy sauce	14	1.46	0.02	2.40	4.7	0.567	0.014	0.036	0	0
Deli meats										
Bologna—beef	89	3.32	8.04	0.56	3.7	0.394	0.016	0.036	0	16.0
Bratwurst	92	4.05	7.90	0.84	13.8	0.292	0.070	0.064	0	17.8
Keilbasa sausage	88	3.76	7.70	0.61	12.0	0.414	0.064	0.061	0	18.5
Knockwurst sausage	87	3.37	7.88	0.05	2.9	0.258	0.097	0.040	0	16.3
Liverwurst	93	4.00	8.10	0.63	7.9	1.810	0.077	0.291	0	44.1
Pepperoni sausage	140	5.94	12.50	0.80	2.8	0.397	0.09	0.07		9.79
Polish sausage	92	3.99	8.13	0.46	3.0	0.409	0.142	0.042	0	20.0
Salami—beef	72	4.17	5.69	0.70	2.47	0.567	0.036	0.073	0	17.3
Salami—pork & beef	72	3.94	5.70	0.64	3.68	0.755	0.068	0.106	0	18.5
Salami—turkey	55	4.62	3.89	0.15	5.47	0.463	0.029	0.075	0	22.9
Salami—dry—beef & pork	120	6.49	9.75	0.74	2.84	0.425	0.170	0.082	0	22.7
Turkey pastrami	37	5.22	1.75	0.43	2.49	0.403	0.022	0.075	0	14.9
Mexican foods										
Beef taco	75.2	4.94	4.80	3.67	30.9	0.469	0.01	0.049	0.407	16.2
Beef enchilada	69.0	3.09	3.26	3.69	60.2	0.418	0.017	0.039	0.465	8.93
Cheese enchilada	78.0	3.12	4.2	3.78	108.0	0.324	0.015	0.050	0.465	10.2
Chicken enchilada	63.6	3.38	2.48	3.69	60.5	0.359	0.018	0.036	0.465	9.10
Corn tortilla, enriched, regular	61.4	1.89	0.964	12.30	39.7	0.567	0.047	0.028	2.27	0
Corn tortilla, enriched, thin	61.0	1.64	0.95	12.30	39.8	0.567	0.046	0.028	2.26	0
Corn tortilla, fried	82.2	2.08	2.84	12.30	39.7	0.567	0.047	0.028	2.27	0
Enchirito	60.4	3.15	2.75	3.31	51.5	0.410	0.015	0.037	0.748	9.18
Flour tortilla	84	2.07	2.15	15.50	17.0	0.440	0.102	0.062	0.800	0
Refried beans, canned	30.3	1.77	0.303	5.24	13.2	0.500	0.014	0.016	2.47	0
Nuts & seeds										
Almonds—dried, chopped	167	5.65	14.80	5.78	75.5	1.040	0.060	0.220	3.36	0
Almonds—whole toasted	167	5.77	14.40	6.48	80.0	1.400	0.037	0.170	3.99	0
Sunflower seeds—dry	162	6.46	14.00	5.32	32.9	1.920	0.650	0.070	1.97	0
Oils & shortening										
Cocoa butter oil	251	0	28.40	0	0	0	0	0	0	0
Corn oil	251	0	28.40	0	0	0.001	0	0	0	0

VARIETY—continued

	kcal	Protein (g)	Fat (g)	CHO (g)	Ca (mg)	Fe (mg)	B₁ (mg)	B₂ (mg)	Fiber (g)	Cholesterol (mg)
Cottonseed oil	251	0	28.40	0	0	0	0	0	0	0
Olive oil	251	0	28.40	0	0.1	0.109	0	0	0	0
Palm oil	251	0	28.40	0	0.1	0.003	0	0	0	0
Palm kernel oil	251	0	28.40	0	0	0	0	0	0	0
Peanut oil	251	0	28.40	0	0	0.008	0	0	0	0
Safflower oil	251	0	28.40	0	0	0	0	0	0	0
Sesame oil	250	0	28.40	0	0	0	0	0	0	0
Soybean oil	251	0	28.40	0	0	0.007	0	0	0	0
Sunflower oil	251	0	28.40	0	0	0	0	0	0	0
Walnut oil	251	0	28.40	0	0	0	0	0	0	0
Wheat germ oil	250	0	28.40	0	0	0	0	0	0	0
Vegetable shortening	251	0	28.40	0	0	0	0	0	0	0
Pasta & noodles										
Spaghetti, cooked firm, hot	41	1.42	0.142	8.53	3.1	0.454	0.051	0.028	0.482	0
Spaghetti, cooked tender, hot	31	1.01	0.111	6.48	3.0	0.405	0.04	0.022	0.425	0
Whole wheat spaghetti, cooked	35	1.53	0.113	7.49	4.3	0.244	0.048	0.020	1.040	0
Spaghetti + sauce + cheese, canned	22	0.68	0.227	4.42	4.5	0.318	0.040	0.032	0.284	0.3
Spaghetti + sauce + cheese, homemade	30	1.02	1.02	4.20	9.07	0.260	0.028	0.020	0.284	0.9
Spaghetti + sauce + meat, canned	30	1.36	1.13	4.42	6.01	0.374	0.017	0.020	0.312	2.6
Spaghetti + sauce + meat, homemade	38	2.17	1.37	4.46	14.20	0.423	0.029	0.034	0.314	10.2
Spaghetti sauce, homemade	23	0.77	1.25	2.95	6.70	0.380	0.026	0.017	0.343	0
Spaghetti sauce, canned	31	0.52	1.35	4.52	7.97	0.184	0.016	0.017	0.343	0
Spaghetti meat sauce,	30	1.03	1.30	3.72	4.95	0.385	0.028	0.022	0.193	2.3
Spaghetti sauce, dry, packet	79	1.70	0.28	18.20	48.20	0.765	NA	0.162	0.057	0
Spaghetti sauce + mushrooms, packet	85	2.84	2.55	13.90	113.00	0.510	NA	0.136	0.085	0
Egg noodles, cooked	35	1.17	0.35	6.60	3.19	0.400	0.039	0.023	0.624	8.9
Chow mein noodles, dry	139	3.72	6.93	16.40	8.82	0.252	0.032	0.019	1.100	3.2
Spinach noodles, dry	108	3.97	1.08	20.20	11.60	1.290	0.278	0.133	1.930	0
Spinach noodles, cooked	32	1.13	0.36	5.97	3.37	0.429	0.043	0.027	0.569	0
Chicken + noodles, recipe	43	2.60	2.13	3.07	3.07	0.278	0.006	0.020	0.142	12.2
Chicken + noodles, frozen	32	2.17	1.30	2.49	15.70	0.393	0.009	0.018	0.022	8.0
Noodles-ramen-beef, cooked	28	0.76	0.94	4.17	NA	NA	NA	NA	0.500	NA
Noodles, ramen, chicken, cooked	25	0.76	0.85	3.63	NA	NA	NA	NA	0.512	NA
Noodles, ramen, oriental	26	0.74	1.07	3.83	NA	NA	NA	NA	0.512	NA
Lasagna, frozen entree	38	2.32	1.71	2.61	34.00	0.343	0.026	0.045	0.194	12.4
Pizza										
Pizza—cheese	69	3.54	2.13	9.21	52.00	0.378	0.080	0.069	0.510	13.2
Pizza—mozzarella	80	7.60	4.67	0.89	207.00	0.076	0.006	0.097	0	15.0
Pizza—Canadian bacon	52	6.82	2.36	0.38	3.00	0.229	0.231	0.055	0	16.3
Pizza—pepperoni	140	5.94	12.50	0.81	2.80	0.397	0.090	0.070	0	9.8
Pizza—onion	10	0.34	0.07	2.07	7.10	1.105	0.017	0.003	0.450	0
Popcorn										
Popcorn—plain, air popped	106	3.50	1.42	21.30	3.50	0.709	0.106	0.035	4.600	0
Popcorn—cooked in oil/salted	142	2.32	7.99	15.50	7.70	0.696	0.026	0.052	3.090	0
Popcorn—syrup-coated	109	1.60	0.81	24.30	1.60	0.405	0.106	0.016	0.810	0

VARIETY—continued

	kcal	Protein (g)	Fat (g)	CHO (g)	Ca (mg)	Fe (mg)	B₁ (mg)	B₂ (mg)	Fiber (g)	Cholesterol (mg)
Rice										
Brown—dry	102	2.13	0.54	21.90	9.04	0.510	0.096	0.014	0.965	0
Brown—cooked	34	0.709	0.17	7.23	3.40	0.170	0.026	0.006	0.483	0
White—regular, dry	103	1.90	0.11	22.80	6.74	0.828	0.125	0.009	0.340	0
White—regular, cooked	31	0.567	0.03	6.86	2.84	0.397	0.03	0.003	0.102	0
White—converted, dry	105	2.10	0.15	23.00	17.00	0.828	0.124	0.010	0.624	0
White—converted, cooked	30	0.599	0.02	6.60	5.35	0.227	0.03	0.003	0.180	0
White—instant, dry	106	2.13	0.06	23.40	1.42	1.300	0.125	0.008	0.737	0
White—instant, prepared	31	0.624	0.03	6.86	0.85	0.227	0.037	0.003	0.216	0
Wild—cooked	26	1.02	0.06	5.39	1.42	0.312	0.031	0.045	0.709	0
Rice bran	78	3.77	5.44	14.40	21.50	5.500	0.64	0.070	6.150	0
Rice polish	75	3.43	3.62	16.40	19.40	4.560	0.521	0.051	0.680	0
Salad dressings										
Blue cheese salad dressing	143	1.37	14.80	2.09	22.90	0.057	0.003	0.028	0.020	7.6
Caesar's salad dressing	126	2.66	12.70	0.67	44.40	0.239	0.007	0.025	0.050	33.9
French dressing	150	0.16	16.00	1.81	3.55	0.113	0	0	0.220	0
Italian dressing—low calorie	15	0.02	1.18	1.37	0.59	0.059	0	0	0.080	1.7
Mayonnaise	203	0.31	22.60	0.77	5.67	0.168	0.005	0.012	0	16.8
Imitation mayonnaise	66	0	5.67	3.78	0	0	0	0	0	7.6
Ranch salad dressing	104	0.86	10.70	1.31	28.40	0.075	0.010	0.040	0	11.1
Russian salad dressing	140	0.45	14.50	3.00	5.44	0.174	0.014	0.014	0.080	18.4
1000 island dressing	107	0.26	10.10	4.30	3.52	0.170	0.006	0.009	0.060	7.3
1000 island dressing—low calorie	45	0.22	3.00	4.58	3.12	1.174	0.006	0.008	0.340	3.4
Vinegar & oil dressing	124	0	14.20	0	0	0	0	0	0	0
Salads, prepared										
Chicken salad w/celery	97	3.82	8.90	0.47	5.92	0.239	0.012	0.028	0.110	17.3
Cole slaw	20	0.36	0.74	3.52	12.80	0.166	0.019	0.017	0.570	2.3
Egg salad	68	2.91	6.01	0.45	14.50	0.525	0.019	0.070	0	97.4
Ham salad spread	62	2.46	4.39	3.01	2.24	0.168	0.123	0.034	0.030	10.4
Macaroni salad—no cheese	75	0.54	6.66	3.52	5.50	0.229	0.020	0.014	0.270	4.9
Potato salad w/mayo+eggs	41	0.76	2.32	3.16	5.44	0.185	0.022	0.017	0.420	19.3
Tuna salad	53	4.55	2.53	2.67	4.84	0.282	0.009	0.019	0.340	3.7
Waldorf salad	85	0.72	8.33	2.62	8.82	0.196	0.020	0.013	0.720	4.3
Sandwiches										
Avocado & cheese—white	64	2.02	4.00	5.39	43.1	0.418	0.057	0.059	0.98	4.4
Avocado & cheese—whole wheat	62	2.14	3.97	5.24	37.8	0.477	0.047	0.05	1.76	4.3
BLT—whole wheat	68	2.49	3.77	6.45	11.4	0.570	0.077	0.043	1.55	4.1
BLT—white	70	2.30	3.77	6.70	18.2	0.489	0.092	0.055	0.43	4.2
Grilled cheese—wheat	91	4.22	5.37	7.10	87.4	0.565	0.057	0.076	1.59	11.8
Grilled cheese—part WW	95	4.39	5.82	6.59	103.0	0.526	0.067	0.093	0.61	13.3
Grilled cheese—white	97	4.22	5.79	6.88	103.0	0.441	0.068	0.092	0.30	13.3
Chicken salad—wheat	80	3.00	4.25	8.13	14.7	0.679	0.066	0.046	1.88	6.2
Chicken salad—white	85	2.83	4.63	8.05	22.7	0.552	0.080	0.060	0.39	7.1
Corn dog	84	2.55	5.10	6.97	8.7	0.495	0.072	0.043	0.03	9.5
Corned beef & swiss on rye	83	5.26	4.59	4.90	63.8	0.768	0.044	0.078	0.97	16.4
English muffin (egg/cheese/bacon)	74	3.70	3.70	6.37	40.5	0.637	0.095	0.103	0.32	43.8
Egg salad—wheat	79	2.60	4.56	7.44	17.0	0.744	0.063	0.059	1.67	48.8
Egg salad—soft white	83	2.41	4.90	7.25	24.5	0.636	0.075	0.074	0.30	41.6
Ham—rye bread	59	3.84	2.09	6.10	12.0	0.474	0.182	0.073	1.24	7.1
Ham—whole wheat	59	3.79	2.04	6.84	12.4	0.617	0.164	0.060	1.54	6.1
Ham—soft white	61	3.74	2.08	6.60	18.6	0.504	0.186	0.073	0.29	6.7
Ham & swiss—rye	68	4.65	3.23	5.08	63.5	0.389	0.147	0.079	0.99	10.8
Ham & cheese—wheat	67	4.20	3.23	5.72	40.5	0.527	0.136	0.066	1.27	9.7

VARIETY — continued

	kcal	Protein (g)	Fat (g)	CHO (g)	Ca (mg)	Fe (mg)	B₁ (mg)	B₂ (mg)	Fiber (g)	Cholesterol (mg)
Sandwiches — continued										
Ham & cheese — soft white	69	4.20	3.36	5.43	48.0	0.428	0.152	0.078	0.23	10.6
Ham salad — wheat	75	2.45	4.02	7.83	11.5	0.569	0.104	0.045	1.52	5.6
Ham salad — white	78	2.25	4.26	7.71	17.5	0.454	0.119	0.057	0.28	6.2
Hotdog/frankfurter & bun	87	2.80	5.10	7.04	19.6	0.570	0.095	0.062	0.40	7.6
Patty melt — gound beef/rye	91	5.09	6.07	3.96	36.5	0.533	0.040	0.072	0.81	17.1
Peanut butter & jam — whole wheat	92	3.40	3.85	12.30	15.4	0.751	0.070	0.045	2.29	0
Peanut butter & jam — white	98	3.29	4.14	12.80	23.4	0.632	0.085	0.059	0.85	0
Reuben grilled	58	3.43	3.38	3.53	43.6	0.633	0.030	0.052	0.80	10.4
Roast beef — whole wheat	64	4.00	2.52	6.71	12.3	0.760	0.061	0.054	1.53	6.2
Roast beef — white	67	3.97	2.63	6.46	18.5	0.665	0.072	0.066	0.27	6.9
Tuna salad — wheat	72	3.29	3.29	8.03	13.0	0.667	0.057	0.040	1.74	5.5
Tuna salad — white	76	3.18	3.47	7.92	19.6	0.555	0.068	0.052	0.44	6.2
Turkey — whole wheat	62	4.09	2.33	6.67	11.6	0.554	0.056	0.044	1.53	6.0
Turkey — white	64	4.07	2.42	6.41	17.8	0.435	0.066	0.055	0.27	6.7
Turkey & ham — rye	58	3.74	2.12	6.18	12.3	0.743	0.060	0.079	1.23	8.4
Turkey & ham — whole wheat	59	3.71	2.07	6.90	12.6	0.846	0.060	0.064	1.54	7.2
Turkey & ham — white	60	3.65	2.10	6.67	18.8	0.762	0.071	0.078	0.28	8.0
Turkey & ham & cheese on rye	68	4.22	3.46	5.04	44.2	0.616	0.050	0.083	0.99	12.1
Turkey & ham & cheese — wheat	67	4.14	3.25	5.77	40.5	0.716	0.051	0.070	1.27	10.6
Turkey & ham & cheese — white	69	4.13	3.38	5.48	48.3	0.636	0.059	0.082	0.23	11.6
Sauces										
Bordelaise sauce	24	0.33	1.46	1.10	3.80	0.179	0.008	0.010	0.01	3.8
Hot chili sauce, red pepper	5	0.25	0.17	1.10	2.51	0.137	0.003	0.026	2.29	0
Teriyaki sauce	24	1.69	0.09	4.52	6.30	0.488	0.008	0.020	0	0
Seafood										
Anchovy — raw	37	5.78	1.37	0	41.7	0.921	0.016	0.073	0	19.6
Frog legs — raw meat	21	4.65	0.085	0	5.1	0.539	0.040	0.070	0	14.2
Lobster meat — cooked	28	5.80	0.17	0.36	17.2	0.111	0.020	0.019	0	20.3
Scampi — fried in crumbs	69	6.07	3.49	3.26	19.0	0.357	0.037	0.039	0.04	50.2
Shrimp — boiled	28	5.93	0.31	0	11.0	0.876	0.009	0.009	0	55.3
Squid (calamari) — fried in flour	50	5.10	2.12	2.20	11.0	0.287	0.016	0.130	0	73.7
Soups										
Cream of celery	20	0.38	1.27	2.00	9.04	0.141	0.007	0.011	0.09	3.2
Chicken, chunky	20	1.43	0.75	1.95	2.71	0.195	0.010	0.020	0.03	3.4
Chicken + dumpling	23	1.30	1.28	1.39	3.34	0.144	0.004	0.017	0.10	7.6
Chicken gumbo	13	0.60	0.32	1.89	5.53	0.201	0.006	0.009	0.05	0.9
Chicken-noodle — chunky	14	1.50	0.70	0.24	2.84	0.170	0.009	0.020	0.09	2.1
Chili with beans	32	1.62	1.56	3.38	13.20	0.973	0.014	0.030	0.91	4.8
Clam chowder — New England	19	1.08	0.75	1.90	21.40	0.169	0.008	0.027	0.11	2.5
Minestrone — chunky	15	0.60	0.33	2.45	7.20	0.209	0.006	0.014	0.12	0.6
Cream of mushroom	29	0.46	2.15	2.10	7.23	0.119	0.007	0.019	0.06	0.4
Mushroom — barley	14	0.43	0.51	1.69	2.82	0.113	0.006	0.020	0.17	0
Onion — canned	13	0.87	0.40	1.89	6.10	0.156	0.008	0.006	0.11	0
Oyster stew	14	0.49	0.89	0.94	4.96	0.227	0.005	0.008	0	3.1
Pea — prepared w/milk	27	1.40	0.79	3.59	19.30	0.224	0.017	0.030	0.07	2.0
Cream of potato	17	0.39	0.53	2.59	4.52	0.107	0.008	0.008	0.10	1.5
Split pea + ham	22	1.31	0.47	3.17	3.90	0.253	0.014	0.011	0.19	0.8
Tomato — canned	19	0.47	0.43	3.75	3.05	0.396	0.020	0.011	0.11	0
Tomato-beef-noodle	32	1.00	0.97	4.78	3.95	0.252	0.019	0.020	0.03	0.9
Tomato bisque prepared w/milk	22	0.71	0.75	3.32	21.00	0.099	0.013	0.030	0.01	2.5

VARIETY — continued

	kcal	Protein (g)	Fat (g)	CHO (g)	Ca (mg)	Fe (mg)	B₁ (mg)	B₂ (mg)	Fiber (g)	Cholesterol (mg)
Soups — continued										
Turkey — chunky	16	1.23	0.53	1.69	6.00	0.229	0.010	0.029	0.12	1.1
Turkey noodle	16	0.88	0.45	1.95	2.60	0.212	0.017	0.014	0.03	1.1
Cream vegetable — dry mix	126	2.27	6.84	14.80	1.42	0.539	1.470	0.127	0.22	1.2
Vegetable	16	0.49	0.45	2.77	4.96	0.249	0.012	0.010	0.37	0
Miscellanous										
Garlic cloves	42	1.80	0.14	9.38	51.30	0.482	0.057	0.030	0.47	0
Gelatin salad/dessert	17	0.43	0	3.99	0.50	0.024	0.002	0.002	0.02	0
Quiche lorraine	97	2.09	7.73	4.67	34.00	0.226	0.018	0.052	0.09	45.9
Spinach souffle	45	2.29	3.84	0.59	47.90	0.279	0.019	0.064	0.79	38.4

Part 2
Nutritive Values for Alcoholic and Nonalcoholic Beverages

The nutritive values for alcoholic and nonalcoholic beverages are expressed in 1-ounce (28.4-g) portions. We have also included the nutritive values for the minerals calcium, iron, magnesium, phosphorus, and potassium and the vitamins B$_1$ (thiamine), B$_2$ (riboflavin), niacin, and B$_{12}$ (cobalamin). The alcoholic beverages contain no cholesterol or fat.

ALCOHOLIC BEVERAGES (1 OUNCE)

	kcal	Protein (g)	CHO (g)	Minerals					Vitamins			
				Ca (mg)	Fe (mg)	Mg (mg)	P (mg)	K (mg)	B$_1$ (mg)	B$_2$ (mg)	Niacin (mg)	B$_{12}$ (mg)
Beer, regular	12	0.072	1.1	1.4	0.009	1.83	3.50	7.09	0.002	0.007	0.128	0.005
Beer, light	8	0.057	0.4	1.4	0.011	1.42	3.44	5.13	0.003	0.008	0.111	0.002
Brandy	69	0	10.6	2.5	0.012		1.01	1.01	0.002	0.002	0.004	0
Champagne	22	0.043	0.6	1.6	0.093	2.40	1.90	22.60	0	0.003	0.019	0
Dessert wine, dry	36	0.057	1.2	2.3	0.068	2.55	2.55	26.20	0.005	0.005	0.060	0
Dessert wine, sweet	44	0.057	3.3	2.3	0.057	2.55	2.64	26.20	0.005	0.005	0.060	0
Gin, rum, vodka, scotch, whiskey, 80 proof	64	0	0	0	0.010	0	0	1.01	0	0	0	0
Gin, rum, vodka, scotch, whiskey, 86 proof	71	0	0	0	0.012	0	1.16	0.55	0.002	0.002	0.004	0
Gin, rum, vodka, scotch, whiskey, 90 proof	74	0	0	0	0.010	0	0	0.86	0	0	0	0
Sherry, dry	28	0.024	0.3	2.1	0.052	1.96	2.60	17.80	0.002	0.002	0.024	0
Sherry, medium	40	0.066	2.3	2.3	0.071	2.27	1.89	23.60	0.002	0.008	0.035	0
Vermouth, dry	34	0.028	1.6	2.0	0.096	1.42	1.89	11.30	NA	NA	0.011	0
Vermouth, sweet	44	0.014	4.5	1.7	0.099	1.13	1.65	8.50	NA	NA	0.011	0
Wine, dry white	19	0.029	0.2	2.6	0.093	2.62	1.67	17.40	0	0.001	0.019	0
Wine, medium white	19	0.028	0.2	2.5	0.085	3.03	3.84	22.60	0.001	0.001	0.019	0
Wine, red	20	0.055	0.5	2.2	0.122	3.60	3.84	31.50	0.001	0.008	0.023	0.004
Wine, rosé	20	0.055	0.4	2.4	0.108	2.74	4.08	28.10	0.001	0.004	0.020	0.002
Creme de menthe	105	0	11.8	0	0.023	0	0	0	0	0	0.001	0
Bloody Mary	22	0.153	0.9	1.9	0.105	2.10	4.02	41.40	0.010	0.006	0.123	0
Bourbon and soda	26	0	0	1.0	0.244	0.24	0.48	0.48	0	0	0.005	0
Daiquiri	52	0	1.9	0.9	0.043	0.47	1.89	6.14	0.004	0.	0.012	0
Manhattan	64	0	0.9	0.5	0.025	0.01	1.99	7.46	0.003	0.001	0.026	0
Martini	63	0	0.1	0.4	0.024	0.40	0.81	5.26	0	0	0.004	0
Pina colada	53	0.120	8.0	2.2	0.062		2.01	20.10	0.008	0.004	0.033	0
Screwdriver	23	0.160	2.5	2.1	0.023	2.26	3.86	43.30	0.018	0.004	0.046	0
Tequila	31	0.099	2.4	1.7	0.077	1.98	2.80	29.30	0.010	0.005	0.054	0
Tom collins	16	0.013	0.4	1.3	NA	0.38	0.12	2.30	0	0	0.004	0
Whiskey sour	42	0	3.7	0.3	0.021	0.26	1.60	5.08	0.003	0.002	0.006	0
Coffee + cream liqueur	93	0.784	5.9	4.2	0.036	0.60	13.90	9.05	0	0.016	0.022	0
Coffee liqueur	95	0	13.3	0.5	0.016	0.54	1.64	8.18	0.001	0.003	0.040	0

NONALCOHOLIC BEVERAGES (1 OUNCE)

	kcal	Protein (g)	CHO (g)	Ca (mg)	Fe (mg)	Mg (mg)	P (mg)	K (mg)	B₁ (mg)	B₂ (mg)	Niacin (mg)	B₁₂ (mg)
				Minerals					**Vitamins**			
Hot cocoa with whole milk	25	1.030	2.9	33.8	0.088	6.350	30.60	54.400	0.012	0.049	0.041	0.099
Cocoa mix+water—diet	7	0.561	1.3	13.3	0.110	4.870	19.80	59.800	0.006	0.030	0.024	0
Coffee—brewed	0.2	0.016	0.1	0.5	0.113	1.590	0.32	15.400	0	0.002	0.063	0
Coffee—instant dry powder	1.4	0.016	0.3	0.8	0.019	1.100	2.05	67.70	0	0	0.061	0
Coffee—cappuchino	9.2	0.059	1.6	1.0	0.022	1.330	3.84	17.600	0	0	0.048	0
Coffee—Swiss mocha	7.7	0.078	1.3	1.1	0.036	1.360	4.37	17.900	0	0	0.039	0
Coffee whitener—nondairy,	38.5	0.284	3.2	2.6	0.009	0.060	18.20	54.1	0	0	0	0
liquid powder	155	1.360	15.6	6.3	0.326	1.200	120.00	230.0	0	0.047	0	0
Cola beverage, regular	12	0	3.0	0.7	0.009	0.230	3.52	0.306	0	0	0	0
Diet cola—w/aspartame	0	0	0	1.0	0.009	0.319	2.40	0	0.001	0.007	0	0
Club soda	0	0	0	1.4	0.012	0.319	0	0.479	0	0	0	0
Cream soda	15	0	3.8	1.5	0.015	0.229	0	0.306	0	0	0	0
Diet soda-avg assorted	0	0	0.0	1.1	0.011	0.200	3.03	0.559	0	0	0	0
Egg nog—commercial	38	1.080	3.8	36.8	0.057	5.250	31.00	46.900	0.010	0.054	0.030	0.127
Five Alive citrus	13	0.135	3.1	1.7	0.021	1.950	2.70	34.000	0.015	0.003	0.060	0
Fruit flavored soda pop	13	0	3.2	1.1	0.020	0.305	0.15	1.520	0	0	0.002	0
Fruit punch drink—canned	13	0.015	3.4	2.1	0.058	0.610	0.31	7.160	0.006	0.006	0.006	0
Gatorade	5	0	1.3	2.8	NA	NA	0	2.840	NA	NA	NA	0
Ginger ale	10	0.008	2.5	0.9	0.051	0.232	0.08	0.387	0	0	0	0
Grape soda carbonated	12	0	3.2	0.9	0.024	0.305	0	0.229	0	0	0	0
Kool-Aid w/NutraSweet	0	0	0	0	0	0.028	0	0	0	0	0	0
Kool-Aid w/sugar added	12	0	3.0	0	0	0	0	0	0	0	0	0
Lemon-lime soda	12	0	3.0	0.7	0.019	0.154	0.08	0.308	0	0	0.004	0
Lemonade drink from dry	11	0	2.9	7.6	0.016	0.322	3.65	3.540	0	0	0.004	0
Lemonade frozen conc	51	0.078	13.3	1.9	0.205	1.420	2.46	19.200	0.007	0.027	0.020	0
Limeade frozen conc	53	0.052	14.0	1.4	0.029	7.800	1.69	16.800	0.003	0.003	0.028	0
Chocolate milkshake	36	0.962	5.8	32.0	0.088	4.700	28.90	56.800	0.016	0.069	0.046	0.097
Strawberry milkshake	32	0.952	5.4	32.0	0.030	3.600	28.40	51.700	0.013	0.055	0.050	0.088
Vanilla milkshake	32	0.982	5.1	34.5	0.026	3.500	29.00	49.300	0.013	0.052	0.052	0.101
Orange drink/carbonated	14	0	3.5	1.5	0.018	0.305	0.31	0.686	0	0	0	0
Pepper-type soda	12	0	2.9	0.9	0.010	0.077	3.16	0.154	0	0	0	0
Root beer	12	0.008	3.0	1.5	0.014	0.306	0.15	0.230	0	0	0	0
Pineapple grapefruit drink	13	0.068	3.3	2.0	0.087	1.700	1.59	17.500	0.009	0.005	0.076	0
Pineapple orange drink	14	0.352	3.3	1.5	0.076	1.590	1.13	13.200	0.009	0.005	0.059	0
Tang orange juice crystals	13	0.170	0.06	3.1	4.57	0.002	0.02	0.008	0	0	0	0
Tonic water/Quinine water	10	0	2.5	0.4	0.019	0.077	0	0.077	0	0	0	0
Tea-brewed	0	0.001	0.1	0	0.006	0.796	0.16	10.500	0	0.004	0.012	0
Herbal tea, brewed	0	0	0	0.6	0.022	0.319	0	2.390	0.003	0.001	0	0
Perrier water	0	0	0	3.8	0	0.148	0	0	0	0	0	0
Poland Springs bottled water	0	0	0	0.4	0.001	0.239	0	0	0	0	0	0

Note: Alcoholic beverages contain no fat or cholesterol; light beer contains 0.5 g fiber and regular beer contains 1.2 g fiber per 8 oz. serving. All of the other nonmixed alcoholic beverages have no fiber.

Note: Other nonalcoholic beverages are listed in the sections on fruits and vegetables.

Part 3
Nutritive Values for Specialty and Fast-Food Items

Nutrient information was kindly provided by the manufacturer or its representative, and is reproduced as presented in their literature. Unlike Parts 1 and 2, nutritive values are not given for 1-ounce portions but for the actual amounts of the foods as sold commercially. To make a direct comparison of the kcal values and the various nutrients, we recommend that the weight of the food and its nutrients be expressed relative to 1-ounce (28.4-g) portions.

ARBY'S

Food Item	Serving Size (g)	kcal	Protein (g)	Fat (g)	CHO (g)
Bacon Cheddar Deluxe	225	561	28	34	78
Baked Potato Plain	312	291	8	1	0
Beef 'n Cheddar	190	490	24	21	51
Chicken Breast Sandwich	210	592	28	27	57
Chocolate Shake	300	384	9	11	32
French Fries	71	211	2	8	6
Hot Ham 'n Cheese Sandwich	161	353	26	13	50
Jamocha Shake	305	424	8	10	31
Junior Roast Beef	86	218	12	8	22
King Roast Beef	192	467	27	19	49
Potato Cakes	85	201	2	14	13
Regular Roast Beef	147	353	22	15	32
Super Roast Beef	234	501	25	22	40
Superstuffed Potato Broccoli and Cheddar	340	541	13	22	24
Superstuffed Potato Deluxe	312	648	18	38	72
Superstuffed Potato Mushroom and Cheese	300	506	16	22	21
Superstuffed Potato Taco	425	619	23	27	145
Turkey Deluxe	197	375	24	17	39
Vanilla Shake	250	295	8	10	30

Source: Arby's Inc. Nutritional information provided by Consumer Affairs, Arby's Inc., Atlanta, GA, 1986.

BURGER KING

Food Item	Serving Size (g)	kcal	Pro-tein (g)	Total Carbo-hydrate (g)	Total Fat (g)	Choles-terol (mg)	Sodium (mg)	Dietary Fiber (g)	Percentage of U.S. RDA* Vit A	Vit C	Ca	Fe
Burgers												
Whopper sandwich	270	640	27	45	39	90	870	3	100	15	8	25
Whopper with cheese sandwich	294	730	33	46	46	115	1300	3	15	15	25	25
Double Whopper sandwich	351	870	46	45	56	170	940	3	10	15	8	40
Double Whopper with cheese sandwich	375	960	52	46	63	195	1360	3	15	15	25	40
Whopper Jr. sandwich	168	420	21	29	24	60	570	2	4	8	6	20
Whopper Jr. with cheese sandwich	180	460	23	29	28	75	780	2	8	8	15	20
Hamburger	129	330	20	28	15	55	570	1	2	0	4	15
Cheeseburger	142	380	23	28	19	65	780	1	6	0	15	15
Double cheese-burger	213	600	41	29	36	135	1040	1	8	0	20	25
Double cheese-burger with bacon	221	640	44	29	39	145	1220	1	8	0	20	25
Sandwich/Side Orders												
BK Big Fish sandwich	255	720	25	59	43	60	1090	2	2	2	6	20
BK Broiler Chicken sandwich	248	540	30	41	29	80	480	2	4	10	4	30
Chicken sandwich	229	700	26	54	43	60	1400	2	*	*	10	20
Chicken Tenders (6 piece)	88	250	16	14	12	35	530	2	*	*	*	4
Broiled Chicken salad†	302	200	21	7	10	60	110	3	100	25	15	20
Garden salad†	215	90	6	7	5	15	110	3	110	50	15	6
Side salad†	133	50	3	4	3	5	55	2	50	20	6	2
French fries (medium, salted)	116	400	5	43	20	0	240	3	*	4	*	6
Onion rings	124	310	4	41	14	0	810	5	*	*	*	*
Dutch apple pie	113	310	3	39	15	0	230	2	*	10	*	8
Drinks												
Vanilla shake (medium)	284	310	9	53	7	20	230	1	6	6	30	*
Chocolate shake (medium)	284	310	9	54	7	20	230	3	6	*	20	10
Chocolate shake (medium, syrup added)	341	460	11	87	7	20	300	1	6	6	30	*
Strawberry shake (medium, syrup added)	341	430	9	83	7	20	260	1	6	6	30	*
Coca-Cola Classic (medium)	22 (fl oz)	260	0	70	0	0	@	0	*	*	*	*
Diet Coke (medium)	22 (fl oz)	1	0	<1	0	0	@	0	*	*	*	*
Sprite (medium)	22 (fl oz)	260	0	66	0	0	@	0	*	*	*	*
Tropicana orange juice	311	140	2	33	0	0	0	0	0	100	0	0
Coffee	355	5	0	1	0	0	5	0	*	*	*	*
Milk — 2% low fat	244	120	8	12	5	20	120	0	10	4	30	*

BURGER KING—continued

Food Item	Serving Size (g)	kcal	Protein (g)	Total Carbohydrate (g)	Total Fat (g)	Cholesterol (mg)	Sodium (mg)	Dietary Fiber (g)	Percentage of U.S. RDA*			
									Vit A	Vit C	Ca	Fe
Breakfast												
Croissan'wich with bacon, egg and cheese	118	350	15	18	24	225	790	<1	8	*	15	10
Croissan'wich with sausage, egg and cheese	159	530	20	21	41	255	1000	<1	8	*	15	15
Croissan'wich with ham, egg and cheese	144	350	18	19	22	230	1390	<1	8	*	15	10
French toast sticks	141	500	4	60	27	0	490	1	*	*	6	15
Hash browns	71	220	2	25	12	0	320	2	10	8	*	2
A.M. Express grape jam	12	30	0	7	0	0	0	0	0	0	0	0
A.M. Express strawberry jam	12	30	0	8	0	0	5	0	0	0	0	0
Condiments/ Toppings												
Processed American cheese	25	90	6	0	8	25	420	0	6	*	15	*
Lettuce	21	0	0	0	0	0	0	0	*	*	*	*
Tomato	28	5	0	1	0	0	0	0	2	8	*	*
Onion	14	5	0	1	0	0	0	0	*	*	*	*
Pickles	14	0	0	0	0	0	140	0	0	0	0	0
Ketchup	14	15	0	4	0	0	180	0	4	*	*	*
Mustard	3	0	0	0	0	0	40	0	0	0	0	0
Mayonnaise	28	210	0	,1	23	20	160	0	*	*	*	*
Tartar sauce	28	180	0	0	19	15	220	0	*	*	*	*
Land O' Lakes whipped classic blend	10	65	0	0	7	0	75	0	8	*	*	*
Bull's Eye barbecue sauce	14	20	0	5	0	0	140	0	*	*	*	2
Bacon bits	3	15	1	0	1	5	0	0	*	*	*	*
Croutons	7	30	,1	4	1	0	75	0	*	*	*	*
Burger King Salad Dressings												
Thousand Island dressing	30	140	0	7	12	15	190	<1	30	*	*	*
French dressing	30	140	0	11	10	0	190	0	15	*	*	*
Ranch dressing	30	180	<1	2	19	10	170	<1	*	*	*	*
Bleu cheese dressing	30	160	2	1	16	30	260	<1	*	*	*	*
Reduced calorie light Italian dressing#	30	15	0	3	0.5	0	50	0	*	*	*	*
Dipping Sauces												
A.M. Express dip	28	80	0	21	0	0	20	0	*	*	*	*
Honey dipping sauce	28	90	0	23	0	0	10	0	*	*	*	*
Ranch dipping sauce	28	170	0	2	17	0	200	0	*	*	*	*
Barbecue dipping sauce	28	35	0	9	0	0	400	0	2	2	*	*
Sweet & sour dipping sauce	28	45	0	11	0	0	50	0	*	*	*	*

Source: From Burger King Corporation, Miami, FL, 1996. For additional information, call 1-800-937-1800.
†=Without dressing   @=Depends on the water supply   *=Less than 2% of the U.S. RDA
* = Values for vitamins A and C and minerals calcium and iron represent percent daily values based on a 2,000 calorie diet
#=Regular Italian dressing—150 calories—16 grams (g) Fat.
—=Negligible.

DAIRY QUEEN

Food Item	Serving Size (g)	Description	kcal	Protein (g)	Fat (g)	CHO (g)	Ca (mg)	Fe (mg)	Vitamin A (IU)	Vitamin C (mg)	Vitamin B₁ (mg)	Vitamin B₂ (mg)
Banana Split	383		540	9	11	103	—	—	—	—	—	—
Big Brazier	213	deluxe	470	28	24	36	111	5.2	—	<2.5	0.34	0.37
Big Brazier	184	regular	184	27	23	37	113	5.2	—	<2.0	0.37	0.39
Big Brazier	213	w/cheese	553	32	30	38	268	5.2	495	<2.3	0.34	0.53
Blizzard Banana Split	—	regular	763	—	—	—	—	—	—	—	—	—
Blizzard Banana Split	—	large	1333	—	—	—	—	—	—	—	—	—
Blizzard chocolate sandwich cookies	—	regular	600	—	—	—	—	—	—	—	—	—
Blizzard chocolate sandwich cookies	—	large	1050	—	—	—	—	—	—	—	—	—
Blizzard German chocolate	—	regular	794	—	—	—	—	—	—	—	—	—
Blizzard German chocolate	—	large	1460	—	—	—	—	—	—	—	—	—
Blizzard, Heath	—	regular	824	—	—	—	—	—	—	—	—	—
Blizzard, Heath	—	large	1212	—	—	—	—	—	—	—	—	—
Blizzard, M&M	—	regular	766	—	—	—	—	—	—	—	—	—
Blizzard, M&M	—	large	1154	—	—	—	—	—	—	—	—	—
Brazier cheese dog	113		330	15	19	24	168	1.6	—	—	—	0.18
Brazier chili dog	128		330	13	20	25	86	2.0	—	11.0	0.15	0.23
Brazier dog	99		273	11	15	23	75	1.5	—	11.0	0.12	0.15
Brazier french fries	71	regular	200	2	10	25	tr	0.4	tr	3.6	0.06	tr
Brazier french fries	113	large	320	3	16	40	tr	0.4	tr	4.8	0.09	0.03
Brazier onion rings	85		300	6	17	33	20	0.4	tr	2.4	0.09	tr
Brazier regular	106		260	13	9	28	70	3.5	—	<1.0	0.28	0.26
Brazier w/cheese	121		318	18	14	30	163	3.5	—	<1.2	0.29	0.29
Buster Bar	149		460	10	29	41	—	—	—	—	—	—
Chicken sandwich	220		670	29	41	46	—	—	—	—	—	—
Cone	213	large	340	9	10	57	—	—	—	—	—	—
Cone	142	regular	240	6	7	38	—	—	—	—	—	—
Cone	85	small	140	3	4	22	—	—	—	—	—	—
Dairy Queen Parfait	284		460	10	11	81	300	1.8	400	tr	0.12	0.43
Dilly Bar	85		240	4	15	22	100	0.4	100	tr	0.06	0.17
Dilly Bar	85		210	3	13	21	—	—	—	—	—	—
Dipped, cone	234	large	510	9	24	64	—	—	—	—	—	—
Dipped, cone	156	regular	340	6	16	42	—	—	—	—	—	—
Dipped, cone	92	small	190	3	9	25	—	—	—	—	—	—
Double Delight	255		490	9	20	69	—	—	—	—	—	—
Double hamburger	210		530	36	28	33	—	—	—	—	—	—
Double w/cheese	239		650	43	37	34	—	—	—	—	—	—
Chocolate dipped cone	234	large	450	10	20	58	300	0.4	400	tr	0.12	0.51
Chocolate dipped cone	156	medium	300	7	13	40	200	0.4	300	tr	0.09	0.34
Chocolate dipped cone	78	small	150	3	7	20	100	tr	100	tr	0.03	0.17
Chocolate malt	588	large	840	22	28	125	600	5.4	750	6.0	0.15	0.85
Chocolate malt	418	medium	600	15	20	89	500	3.6	750	3.6	0.12	0.60
Chocolate malt	241	small	340	10	11	51	300	1.8	400	2.4	0.06	0.34
Chocolate sundae	248	large	400	9	9	71	300	1.8	400	tr	0.09	0.43
Chocolate sundae	184	medium	300	6	7	53	200	1.1	300	tr	0.06	0.26

DAIRY QUEEN—continued

Food Item	Serving Size (g)	Description	kcal	Protein (g)	Fat (g)	CHO (g)	Ca (mg)	Fe (mg)	Vitamin A (IU)	Vitamin C (mg)	Vitamin B₁ (mg)	Vitamin B₂ (mg)
Chocolate sundae	106	small	170	4	4	30	100	0.7	100	tr	0.03	0.17
Cone	213	large	340	10	10	52	300	tr	400	tr	0.15	0.43
Cone	142	medium	230	6	7	35	200	tr	300	tr	0.09	0.26
Cone	71	small	110	3	3	18	100	tr	100	tr	0.03	0.14
Float	397		330	6	8	59	200	tr	100	tr	0.12	0.17
Freeze	397		520	11	13	89	300	tr	200	tr	0.15	0.34
Sandwich	60		140	3	4	24	60	0.4	100	tr	0.03	0.14
Fiesta sundae	269		570	9	22	84	200	tr	200	tr	0.23	0.26
Fish sandwich	170		400	20	17	41	60	1.1	tr	tr	0.15	0.26
Fish sandwich w/cheese	177		440	24	21	39	150	0.4	100	tr	0.15	0.26
Float	397		410	5	7	82	—	—	—	—	—	—
Freeze	397		500	9	12	89	—	—	—	—	—	—
French fries	71	regular	200	2	10	25	—	—	—	—	—	—
French fries	113	large	320	3	16	40	—	—	—	—	—	—
Frozen dessert	113		180	4	6	27	—	—	—	—	—	—
Hot dog	100		280	11	16	21	—	—	—	—	—	—
Hot dog w/cheese	114		330	15	21	21	—	—	—	—	—	—
Hot dog w/chili	128		320	13	20	23	—	—	—	—	—	—
Hot Fudge brownie delight	266		600	9	25	85	—	—	—	—	—	—
Malt	588	large	1060	20	25	187	—	—	—	—	—	—
Malt	418	regular	760	14	18	134	—	—	—	—	—	—
Malt	291	small	520	10	13	91	—	—	—	—	—	—
Mr. Misty	439	large	340	0	0	84	—	—	—	—	—	—
Mr. Misty	330	regular	250	0	0	63	—	—	—	—	—	—
Mr. Misty	248	small	190	0	0	48	—	—	—	—	—	—
Mr. Misty float	404		440	6	8	85	200	tr	120	tr	0.12	0.17
Mr. Misty float	411		390	5	7	74	—	—	—	—	—	—
Mr. Misty freeze	411		500	9	12	91	—	—	—	—	—	—
Mr. Misty Kiss	89		70	0	0	17	—	—	—	—	—	—
Onion rings	85		280	4	16	31	—	—	—	—	—	—
Parfait	283		430	8	8	76	—	—	—	—	—	—
Peanut Buster Parfait	305		740	16	34	94	—	—	—	—	—	—
Shake	588	large	990	19	26	168	—	—	—	—	—	—
Shake	418	regular	710	14	19	120	—	—	—	—	—	—
Shake	291	small	490	10	13	82	—	—	—	—	—	—
Single hamburger	148		360	21	16	33	—	—	—	—	—	—
Single w/cheese	162		410	24	20	33	—	—	—	—	—	—
Strawberry shortcake	312		540	10	11	100	—	—	—	—	—	—
Sundae	248	large	440	8	10	78	—	—	—	—	—	—
Sundae	177	regular	310	5	8	56	—	—	—	—	—	—
Sundae	106	small	190	3	4	33	—	—	—	—	—	—
Super Brazier	298		783	53	48	35	282	7.3	—	,3.2	0.39	0.69
Super Brazier chili dog	210		555	23	33	42	158	4.0	—	18.0	0.42	0.48
Super Brazier dog	182		518	20	30	41	158	4.3	tr	14.0	0.42	0.44
Super Brazier dog w/cheese	203		593	26	36	43	297	4.4	—	14.0	0.43	0.48
Super hot dog	175		520	17	27	44	—	—	—	—	—	—
Super hot dog w/cheese	196		580	22	34	45	—	—	—	—	—	—
Super hot dog w/chili	218		570	21	32	47	—	—	—	—	—	—
Triple hamburger	272		710	51	45	33	—	—	—	—	—	—
Triple w/cheese	301		820	58	50	34	—	—	—	—	—	—

Source: International Dairy Queen, Inc., Minneapolis, MN, 1982. Nutritional information reviewed and edited by Dr. David J. Aulik in cooperation with Raltech Scientific Services.

JACK IN THE BOX

Menu Item	Serving Size (g)	Calories (per serving)	Protein (g)	Fat (g)	Carbo-hydrates (g)	Percentage of U.S. RDA			
						Calcium	Iron	Vitamin A	Vitamin C
Sandwiches									
Beef Gyro	260	620	27	32	55	8	30	4	15
Chicken Fajita Pita	189	290	24	8	29	25	15	10	10
Chicken sandwich	160	400	20	18	38	15	10	4	0
Chicken supreme	245	620	25	36	48	20	15	10	4
Country fried steak sandwich	153	450	14	25	42	6	15	2	8
Fish supreme	245	590	22	32	51	20	20	10	8
Grilled chicken fillet	211	430	29	19	36	15	35	6	10
Smoked chicken cheddar & bacon	223	540	30	30	37	30	15	10	15
Sourdough ranch chicken sandwich	225	490	29	21	45	—	10	—	0
Spicy crispy chicken sandwich	224	560	24	27	55	10	15	4	8
Hamburgers									
Hamburger	97	280	13	11	31	10	15	2	2
Cheeseburger	110	330	16	15	32	20	15	6	2
Double cheeseburger	152	450	24	24	35	25	20	10	0
Jumbo Jack	229	560	26	32	41	10	25	4	10
Jumbo Jack with cheese	242	610	29	36	41	20	30	6	10
Bacon bacon cheeseburger	242	710	35	45	41	25	30	8	15
Grilled sourdough burger	223	670	32	43	39	20	25	15	10
Ultimate cheeseburger	280	830	47	57	33	30	35	15	0
1/4 lb. burger	172	510	26	27	39	15	20	6	0
The Colossus burger	272	940	52	60	48	30	35	10	6
Teriyaki Bowls									
Chicken Teriyaki bowl	440	580	28	1.5	115	10	10	110	15
Beef Teriyaki bowl	440	580	28	3	124	15	25	100	10
Soy sauce	9	5	**	0	**	0	0	0	0
Breakfast									
Breakfast Jack	121	300	18	12	30	20	15	8	15
Pancake platter	231	610	15	22	87	10	10	8	10
Sausage crescent	156	580	22	43	28	15	15	10	0
Scrambled egg platter	213	560	18	32	50	15	25	15	15
Scrambled egg pocket	183	430	29	21	31	20	20	20	0
Sourdough breakfast sandwich	147	380	21	20	31	25	20	15	15
Supreme crescent	153	530	23	33	34	15	20	15	20
Hash browns	57	160	1	11	14	0	2	0	10
Country Crock Spread	5	25	0	3	0	0	2	4	0
Grape jelly	14	40	0	0	9	0	0	0	0
Pancake syrup	42	120	0	0	30	0	0	0	0
Salads									
Chef salad	324	320	19	30	9	35	15	90	35
Taco salad	397	470	23	34	30	40	25	30	15
Side salad	111	50	3	7	**	6	0	60	15
Bleu cheese dressing	70	260	22	**	14	0	0	0	0
Buttermilk house dressing	70	360	36	**	8	0	0	0	0
Low calorie Italian dressing	70	25	2	**	2	0	0	0	0
Thousand Island dressing	70	310	30	**	12	0	0	0	0
Mexican Food									
Taco	78	190	11	7	15	10	6	0	0
Super taco	126	280	17	12	22	15	10	0	4
Guacamole	25	50	4	**	3	*	*	2	70
Salsa	28	10	0	0	2	*	*	2	*
Sides & Desserts									
Seasoned curly fries	109	360	5	20	39	2	8	0	8
Small french fries	68	220	3	11	28	0	4	0	30
Regular french fries	109	350	4	17	45	0	6	0	40
Jumbo fries	123	400	5	19	51	0	8	0	45
Onion rings	103	380	5	23	38	2	10	0	4
Sesame breadsticks	16	70	2	2	12	0	0	0	0
Tortilla chips	28	140	2	6	18	0	0	0	0
Hot apple turnover	110	350	3	19	48	0	10	0	15
Cheesecake	99	310	8	18	29	10	2	0	0
Chocolate chip cookie dough cheesecake	102	360	7	18	44	15	6	6	0
Cinnamon Churritos	75	330	3	21	34	2	30	0	0

JACK IN THE BOX—continued

Menu Item	Serving Size (g)	Calories (per serving)	Protein (g)	Fat (g)	Carbo-hydrates (g)	Percentage of U.S. RDA			
						Calcium	Iron	Vitamin A	Vitamin C
Finger Foods									
Egg rolls-3 piece	165	440	3	24	54	8	15	0	6
Egg rolls-5 piece	285	750	5	41	92	15	20	0	10
Chicken strips (breaded)-4 piece	112	290	25	13	18	0	4	0	0
Chicken strips (breaded)-6 piece	177	450	39	20	28	0	6	0	0
Chicken Taquitos-5 piece	136	350	19	15	34	15	10	4	2
Chicken Taquitos-8 piece	218	560	30	25	54	20	15	8	4
Barbeque sauce	28	45	1	0	11	0	0	0	0
Buttermilk house sauce	25	130	**	13	3	0	*	0	2
Hot sauce	13	5	**	0	1	0	0	0	0
Sweet & sour sauce	28	40	**	0	11	0	0	0	0
Drinks									
Orange juice	183	80	1	0	20	2	2	8	160
Lowfat milk (2%)	244	120	8	5	12	30	0	10	4
Vanilla milk shake (regular)	304	350	9	7	62	30	0	0	0
Chocolate milk shake (regular)	322	330	11	7	59	35	4	0	0
Strawberry milk shake (regular)	298	330	9	7	60	30	0	0	0
Iced tea (small)	16	0	0	0	0	0	0	0	0
Coffee (small)	8	5	0	0	1	0	0	0	0

Source: Jack In The Box; nutritional information provided by Foodmaker, Inc., San Diego, CA.

KENTUCKY FRIED CHICKEN

Food Item	Serving Size (g)	kcal	Protein (g)	CHO (g)	Fat (g)	Cholesterol (mg)	Vit A (mg)	Vit C (mg)	Thia (mg)	Ribo (mg)	Nia (mg)	Ca (mg)	Fe (mg)
Original Recipe Chicken													
Wing	55	178	12.2	6.0	11.7	64	<100	<1.0	0.03	0.08	3.7	47.9	1.2
Side breast	90	267	18.8	10.8	16.5	77	<100	<1.0	0.06	0.13	6.9	68.0	1.2
Center breast	115	283	27.5	8.8	15.3	93	<100	<1.0	0.09	0.17	11.5	68.0	1.0
Drumstick	57	146	13.1	4.2	8.5	67	<100	<1.0	0.05	0.12	3.2	21.2	1.1
Thigh	104	294	17.9	11.1	19.7	123	104	<1.0	0.08	0.30	5.5	65.1	1.3
Extra Tasty Crispy Chicken													
Wing	65	254	12.4	9.3	18.6	67	<100	<1.0	0.04	0.06	3.3	17.8	0.6
Side breast	110	343	21.7	14.0	22.3	81	<100	<1.0	0.09	0.10	8.5	30.4	0.8
Center breast	135	342	33.0	11.7	19.7	114	<100	<1.0	0.11	0.13	13.1	33.3	0.8
Drumstick	69	204	13.6	6.1	13.9	71	<100	<1.0	0.06	0.12	3.7	12.9	0.7
Thigh	119	406	20.0	14.4	29.8	129	131	<1.0	0.10	0.21	6.5	49.0	1.2
Kentucky Nuggets	16	46	2.8	2.2	2.9	11.9	<100	<1.0	<0.01	0.03	1.00	2.4	0.1
Barbeque sauce	28.3	35	0.3	7.1	0.6	<1.0	<370	<1.0	<0.01	0.01	0.19	6.1	0.2
Sweet 'n sour	28.3	58	0.1	13.0	0.6	<1.0	<100	<1.0	<.01	0.02	0.04	4.7	0.2
Honey	14.2	49	0.0	12.1	<0.01	<1.0	<100	<1.0	<0.01	0.00	0.04	0.6	0.1
Mustard	28.3	36	0.9	6.1	0.9	<1.0	<100	<1.0	<0.01	0.01	0.16	10.2	0.3
Chicken Littles	47	169	5.7	13.8	10.1	18	<100	<1.0	0.16	0.12	2.2	22.6	1.7
Buttermilk biscuits	65	235	4.5	28.0	11.7	1	<100	<1.0	0.24	0.19	2.6	95.0	1.6
Mashed potatoes w/gravy	98	71	2.4	11.7	1.6	<1	<100	<1.0	<0.01	0.04	1.2	21.8	0.04
French fries	77	244	3.2	31.1	11.9	2	<100	15.7	0.15	0.05	2.0	12.5	0.06
Corn on the cob	143	176	5.1	31.9	3.1	<1	272	2.3	0.14	0.11	1.8	7.2	0.08
Cole slaw	91	119	1.5	13.2	6.6	5	310	21.5	0.03	0.03	0.2	32.8	0.02
Colonel's Chicken sandwich	166	482	20.8	38.6	27.3	47	<100	<1.0	0.38	0.27	11.1	46.1	1.3

Source: Public Affairs Department, KFC Corporation, Louisville, KY.

LONG JOHN SILVER'S

Food Item	Serving Size (g)	Description	kcal	Protein (g)	Fat (g)	CHO (g)
3 Pc. Nugget dinner		6 chicken nuggets, Fryes, slaw	699	23	45	54
Apple pie	113		280	2	11	43
Barbecue sauce	34		45	0	0	11
Battered shrimp dinner		6 battered shrimp, Fryes, slaw	771	17	45	60
Bleu cheese dressing	45		225	4	23	3
Breaded clams			465	13	25	46
Breaded fish sandwich platter		Fish sandwich, Fryes, slaw	835	30	42	84
Breaded oysters		6 pieces	460	14	19	58
Breaded shrimp platter		Breaded shrimp, Fryes, slaw, 2 hush puppies	962	20	57	93
Cherry pie	113		294	3	11	46
Chicken planks		4 pieces	458	27	23	35
Clam chowder	187		128	7	5	15
Clam dinner		Clams, Fryes, slaw	955	22	58	100
Cole slaw			138	1	8	16
Cole slaw, drained on fork	98		182	1	15	11
Combo salad		4.25 oz. seafood salad, 2 oz salad shrimp, 6 oz lettuce, 2.4 oz tomato, 1 pkg crackers	397	27	29	21
Corn on the cob	150	1 ear	176	5	4	29
Fish & Chicken		1 fish, 2 tender chicken planks, Fryes, slaw	935	36	55	73
Fish & Fryes		3 fish, Fryes	853	43	48	64
Fish & Fryes		2 pc fish, Fryes	651	30	36	53
Fish & More		2 fish, Fryes, slaw, 2 hush puppies	978	34	58	92
Fish w/batter		2 pieces	319	19	19	19
Fish w/batter		3 pieces	477	28	28	28
Four nuggets and Fryes			427	16	24	39
Fryes	85		247	4	12	31
Fryes			275	4	15	32
Honey-Mustard sauce	35		56	—	—	14
Hush Puppies	47	2 pieces	145	3	7	18
Hush Puppies		3 pieces	158	1	7	20
Kitchen—breaded fish (three piece dinner)		3 kitchen breaded fish, Fryes, slaw, 2 hush puppies	940	35	52	84
Kitchen—breaded (two piece dinner)		2 kitchen-breaded fish, Fryes, slaw, 2 hush puppies	818	26	46	76
Lemon Meringue pie	99		200	2	6	37
Ocean chef salad		6 oz lettuce, 1.25 oz shrimp, 2 oz seafood blend, 2 tomato wedges, 3/4 oz cheese	229	27	8	13
Ocean scallops		6 pieces	257	10	12	27
One fish and Fryes			449	16	24	42
One fish, two nuggets, and Fryes			539	23	30	46

LONG JOHN SILVER'S — continued

Food Item	Serving Size (g)	Description	kcal	Protein (g)	Fat (g)	CHO (g)
Oyster dinner		6 oysters, Fryes, slaw	789	17	45	78
Pecan pie	113		446	5	22	59
Peg leg w/batter		5 pieces	514	25	33	30
Pumpkin pie	113		251	4	11	34
Reduced calorie Italian dressing	49		20	0	1	3
Scallop dinner		6 scallops, Fryes,	747	17	45	66
Sea salad dressing	45		220	4	21	5
Seafood platter		1 fish, 2 battered shrimp, 2 scallops, Fryes, slaw	976	29	58	85
Seafood salad		5.6 oz seafood salad, 6 oz lettuce, 2.4 oz tomato	426	19	30	22
Shrimp & Fish dinner		1 fish, 3 battered shrimp, Fryes, slaw, 2 hush puppies	917	27	55	80
Shrimp salad shrimp, 6 oz lettuce, 2.4 oz tomato		4.5 oz salad	203	28	3	16
Shrimp w/batter		5 pieces	269	9	13	31
Sweet-n-sour sauce	30		—	—	—	—
Tartar sauce	30		117	—	11	5
Tender chicken plank dinner		3 chicken planks, Fryes, slaw	885	32	51	72
Tender chicken plank dinner		4 chicken planks, Fryes, slaw	1037	41	59	82
Thousand Island dressing	48		223	—	22	8
Three piece fish dinner		3 fish, Fryes, slaw, 2 hush puppies	1180	47	70	93
Treasure chest		2 pc fish, 2 peg legs	467	25	29	27
Two planks and Fryes			551	22	28	51

Source: Long John Silver's Seafood Shoppes; sampling and nutrient analysis conducted independently by the Department of Nutrition and Food Science, University of Kentucky, April 10, 1986.

MCDONALD'S

Menu Item	Serving Size	Calories	Protein (g)	Carbo-hydrates (g)	Total Fat (g)	Choles-terol (mg)	Sodium (mg)	Dietary Fiber (g)	Percentage of U.S. RDA			
									Vitamin A	Vitamin C	Calcium	Iron
Sandwiches												
Hamburger	160 g	270	12	34	10	30	520	2	2	4	15	15
Cheeseburger	120 g	320	15	35	14	45	750	2	6	4	15	15
Quarter Pounder	172 g	420	23	37	21	70	690	2	4	4	15	25
Quarter Pounder with cheese	200 g	530	28	38	30	95	1160	2	10	4	15	25
McLean deluxe	214 g	350	24	38	12	60	800	3	8	15	15	25
McLean deluxe with cheese	228 g	400	27	39	17	70	1040	3	10	15	15	25
Big Mac	216 g	530	25	47	28	80	960	3	6	4	20	25
Filet-O-Fish	143 g	360	14	40	16	35	690	2	2	*	10	10
McGrilled chicken Classic	189 g	260	24	33	4	45	500	2	4	8	10	10
McChicken sandwich	190 g	510	17	44	30	50	820	2	2	2	15	15
French Fries												
Small French fries	68 g	210	3	26	10	0	135	2	*	15	*	2
Large French fries	147 g	450	6	57	22	0	290	5	*	30	2	6
Super Size French fries	176 g	540	8	68	26	0	350	6	*	35	2	8
Chicken McNuggets/Sauce												
Chicken McNuggets (4 piece)	73 g	200	12	10	12	40	350	0	*	*	*	4
Chicken McNuggets (6 piece)	109 g	300	19	16	18	65	530	0	*	*	2	6
Chicken McNuggets (9 piece)	165 g	450	28	24	27	95	800	0	*	*	2	8
Hot Mustard (1 pkg)	28 g	60	1	7	3.5	5	240	<1	*	*	*	4
Barbeque sauce (1 pkg)	28 g	45	0	10	0	0	250	0	*	6	*	*
Sweet 'n sour sauce (1 pkg)	28 g	50	0	11	0	0	140	0	6	*	*	*
Honey (1 pkg)	14 g	45	0	12	0	0	0	0	*	*	*	*
Honey mustard (1 pkg)	14 g	50	0	3	4.5	10	85	0	*	*	*	*
Salads												
Chef salad	313 g	210	19	9	11	180	730	2	90	35	16	10
Fajita chicken salad	285 g	160	20	9	6	65	400	3	160	50	4	10
Garden salad	234 g	80	6	7	4	140	60	2	60	35	6	8
Side salad	139 g	45	3	4	2	70	35	1	50	20	4	4
Croutons (1 pkg)	11 g	50	1	7	1.5	0	125	0	*	*	2	2
Bacon bits (1 pkg)	3 g	15	1	0	1	5	90	0	*	*	*	*
Salad Dressings												
Bleu cheese (1 pkg)	60 g	190	2	8	17	30	650	0	2	*	6	2
Ranch (1 pkg)	60 g	230	1	10	21	20	550	0	*	2	4	*
1000 Island	63 g	190	1	16	13	25	510	1	2	2	2	2
Lite Vinaigrette (1 pkg)	62 g	50	0	9	2	0	240	0	6	6	*	*
Red French reduced calorie (1 pkg)	68 g	160	0	23	8	0	490	0	4	6	*	*
Breakfast												
Egg McMuffin	137 g	290	17	27	13	235	730	1	10	2	15	15
Sausage McMuffin	112 g	360	13	26	23	45	750	1	4	*	15	10
Sausage McMuffin with Egg	163 g	440	19	27	29	255	820	1	10	*	15	15
English muffin	55 g	140	4	25	2	0	220	1	*	*	10	8
Sausage biscuit	119 g	430	10	32	29	35	1130	1	*	*	8	15
Sausage biscuit with egg	170 g	520	16	33	35	245	1220	1	6	*	10	15
Bacon, egg & cheese biscuit	152 g	450	17	33	27	240	1340	1	10	*	10	15
Biscuit	76 g	260	4	32	13	0	840	1	*	*	6	10
Sausage	43 g	170	6	0	16	35	290	0	*	*	*	2
Scrambled eggs (2)	102 g	170	13	1	12	425	190	0	10	*	6	6
Hash browns	53 g	130	1	14	8	0	330	1	*	4	*	2
Hotcakes (plain)	150 g	310	9	53	7	15	610	2	*	*	10	15

MCDONALD'S—continued

Menu Item	Serving Size	Calories	Protein (g)	Carbohydrates (g)	Total Fat (g)	Cholesterol (mg)	Sodium (mg)	Dietary Fiber (g)	Percentage of U.S. RDA Vitamin A	Vitamin C	Calcium	Iron
Breakfast—cont'd												
Hotcakes (margarine 2 pats & syrup)	222 g	580	9	100	16	15	760	2	8	*	10	15
Cheerios (1 pkg)	19 g	70	2	15	1	0	180	2	15	15	2	25
Wheaties (1 pkg)	23 g	80	2	18	0.5	0	160	2	15	15	4	30
Muffins/Danish												
Fat free apple bran muffin	70 g	170	4	38	0	0	200	1	*	*	4	6
Apple danish	105 g	360	5	51	16	40	290	1	10	*	8	6
Cheese danish	105 g	410	7	47	22	70	340	0	15	*	8	6
Cinnamon Raisin danish	105 g	430	5	56	22	50	280	1	10	*	10	8
Raspberry danish	105 g	400	5	58	16	45	300	1	10	*	8	6
Desserts/Shakes												
Vanilla lowfat frozen yogurt cone	90 g	120	4	24	0.5	5	85	0	*	2	15	2
Strawberry lowfat frozen yogurt sundae	178 g	240	6	51	1	5	115	1	*	2	20	2
Hot caramel lowfat frozen yogurt sundae	182 g	310	7	63	3	5	200	1	2	2	25	*
Hot fudge sundae	179 g	290	8	54	5	5	190	2	*	2	25	4
Nuts (sundaes)	7 g	40	2	2	3.5	0	0	0	*	*	*	*
Baked apple pie	77 g	260	3	34	13	0	200	<1	*	40	2	6
McDonaldland cookies (1 pkg)	56 g	260	4	41	9	0	270	1	*	*	*	10
Vanilla shake—small	414 mL	340	11	62	5	25	220	0	4	6	40	2
Chocolate shake—small	414 mL	340	12	64	5	25	300	1	4	6	45	4
Strawberry shake—small	414 mL	340	12	63	5	25	220	0	4	10	40	4
Milk/Juices												
1% lowfat milk (8 fl oz)	1 crtn.	100	8	13	2.5	10	115	0	10	4	30	*
Orange juice (6 fl oz)	177 mL	80	1	20	0	0	20	0	2	90	2	2
Apple juice (6 fl oz)	177 mL	80	0	20	0	0	0	0	*	*	*	2

Source: McDonald's Corporation. Oak Brook, IL 60521. (708) 575-3663. 1996.
*=less than 2% of the U.S. RDA

PIZZ HUT

Food Item (1 slice)	kcal	Protein (g)	Fat (g)	CHO (g)	Ca (mg)	Fe (mg)	Vitamin A (IU)	Vitamin C (mg)	Vitamin B_1 (mg)	Vitamin B_2 (mg)
Thick 'n' chewy, beef	620	38	20	73	400	7.2	750	<1.2	0.68	0.60
Thick 'n' chewy, cheese	560	34	14	71	500	5.4	1000	<1.2	0.68	0.68
Thick 'n' chewy, pepperoni	560	31	18	68	400	5.4	1250	3.6	0.68	0.68
Thick 'n' chewy, pork	640	36	23	71	400	7.2	750	1.2	0.90	0.77
Thick 'n' chewy, supreme	640	36	22	74	400	7.2	1000	9.0	0.75	0.85
Thin 'n' crispy, beef	490	29	19	51	350	6.3	750	<1.2	0.30	0.60
Thin 'n' crispy, cheese	450	25	15	54	450	4.5	750	<1.2	0.30	0.51
Thin 'n' crispy, pepperoni	430	23	17	45	300	4.5	1000	<1.2	0.30	0.51
Thin 'n' crispy, pork	520	27	23	51	350	6.3	1000	<1.2	0.38	0.68
Thin 'n' crispy, supreme	510	27	21	51	350	7.2	1250	2.4	0.38	0.68

Source: Research 900 and Pizza Hut, Inc., Wichita, KS.

ROY ROGERS

Food Item	Serving Size (g)	kcal	Protein (g)	Fat (g)	CHO (g)
Apple danish	71	249	4.5	11.6	31.6
Bacon cheeseburger	180	581	32.3	39.2	25.0
Biscuit	63	231	4.4	12.1	26.2
Breakfast crescent sandwich	127	401	13.3	27.3	25.3
Breakfast crescent sandwich w/bacon	133	431	15.4	29.7	25.5
Breakfast crescent sandwich w/ham	165	557	19.8	41.7	25.3
Breakfast crescent sandwich w/sausage	162	449	19.9	29.4	25.9
Breast & wing	196	604	43.5	36.5	25.4
Brownie	64	264	3.3	11.4	37.3
Caramel sundae	145	293	7.0	8.5	51.5
Cheese danish	71	254	4.9	12.2	31.4
Cheeseburger	173	563	29.5	37.3	27.4
Cherry danish	71	271	4.4	14.4	31.7
Chicken breast	144	412	33.0	23.7	16.9
Chocolate shake	319	358	7.9	10.2	61.3
Cole slaw	99	110	1.0	6.9	11.0
Crescent roll	70	287	4.7	17.7	27.2
Egg and biscuit platter	165	394	16.9	26.5	21.9
Egg and biscuit platter w/bacon	173	435	19.7	29.6	22.1
Egg and biscuit platter w/ham	200	442	23.5	28.6	22.5
Egg and biscuit platter w/sausage	203	550	23.4	40.9	21.9
French fries	85	268	3.9	13.5	32.0
Hamburger	143	456	23.8	28.3	65.6
Hot chocolate	8 oz	123	3.0	2.0	22.0
Hot fudge sundae	151	337	6.5	12.5	53.3
Hot topped potato plain	227	211	5.9	0.2	47.9
Hot topped potato w/bacon 'n cheese	248	397	17.1	21.7	33.3
Hot topped potato w/broccoli 'n cheese	312	376	13.7	18.1	39.6
Hot topped potato w/oleo	236	274	5.9	7.3	47.9
Hot topped potato w/sour cream 'n chives	297	408	7.3	20.9	47.6
Hot topped potato w/taco beef 'n cheese	359	463	21.8	21.8	45.0
Large fries	113	357	5.3	18.4	42.7
Large roast beef	182	360	33.9	11.9	29.6
Large roast beef w/cheese	211	467	39.6	20.9	30.3
Leg	53	140	11.5	8.0	5.5
Macaroni	100	186	3.1	10.7	19.4
Milk	8 oz	150	8.0	8.2	11.4
Orange juice	8 oz	99	1.5	0.2	22.8
Orange juice	8 oz	136	2.0	0.3	31.3
Pancake platter (w/syrup, butter)	165	452	7.7	15.2	71.8
Pancake platter (w/syrup, butter) w/bacon	173	493	10.4	18.3	72.0
Pancake platter (w/syrup, butter) w/ham	200	506	14.3	17.3	72.4
Pancake platter (w/syrup, butter) w/sausage	203	608	14.2	29.6	71.8
Potato salad	100	107	2.0	6.1	10.9
Roast beef sandwich	154	317	27.2	10.2	29.1
Roast beef sandwich w/cheese	182	424	32.9	19.2	29.9
RR bar burger	208	611	36.1	39.4	28.0
Salad bar 1,000 Island	2 T	160	NA	16.0	4.0
Salad bar bacon 'n tomato	2 T	136	NA	12.0	6.0
Salad bar bacon bits	1 T	24	4.0	1.0	38.0
Salad bar blue cheese dressing	2 T	150	2.0	16.0	2.0
Salad bar cheddar cheese	1/4 cup	112	5.8	9.0	0.8
Salad bar Chinese noodles	1/4 cup	55	1.5	2.8	6.5
Salad bar chopped eggs	2 T	55	4.0	4.0	0.7
Salad bar croutons	2 T	132	5.5	0	31.0
Salad bar cucumbers	5–6 slices	4	NA	0	1.0
Salad bar green peas	1/4 cup	7	0.5	0	1.2
Salad bar green peppers	2 T	4	0.3	0	1.0
Salad bar lettuce	1 cup	10	NA	0	4.0
Salad bar lo-cal Italian	2 T	70	NA	6.0	2.0
Salad bar macaroni salad	2 T	60	1.0	3.6	6.2
Salad bar mushrooms	1/4 cup	5	0.5	0	0.7
Salad bar potato salad	2 T	50	1.0	3.0	5.5
Salad bar ranch	2 T	155	NA	14.0	4.0
Salad bar shredded carrots	1/4 cup	12	0.6	0	24.0
Salad bar sliced beets	1/4 cup	16	0.5	0	3.8
Salad bar sunflower seeds	2 T	101	4.0	9.0	5.0
Salad bar tomatoes	3 slices	20	0.8	0	4.8

ROY ROGERS—continued

Food Item	Serving Size (g)	kcal	Protein (g)	Fat (g)	CHO (g)
Strawberry shake	312	315	7.6	10.2	49.4
Strawberry shortcake	205	447	10.1	19.2	59.3
Strawberry sundae	142	216	5.7	7.1	33.1
Thigh	98	296	18.4	19.5	11.7
Thigh & leg	151	436	29.9	27.5	17.2
Vanilla shake	306	306	8.0	10.7	45.0
Wing	52	192	10.5	12.8	8.5

Source: Roy Rogers Restaurants, Marriott Corporation, Washington, DC. Nutritional data furnished by Lancaster Laboratories, 1985.

TACO BELL

Food Item	kcal	Protein (g)	Fat (g)	Calories From Fat	Saturated Fat	Cholesterol (mg)	Na (mg)	Percent U.S. RDA Ca	Percent U.S. RDA Fe
Tacos & Tostadas									
Chicken soft taco	223	14	10	90	4	58	553	6	8
Soft taco	223	12	11	100	5	32	539	5	10
Soft taco supreme	268	13	15	140	8	47	551	8	11
Steak soft taco	217	12	9	80	4	31	569	5	6
Taco	180	10	11	100	5	32	276	8	5
Taco supreme	225	11	15	130	7	47	287	10	6
Tostada	242	9	11	100	4	14	593	17	8
Burritos									
7 layer burrito	485	15	21	190	8	28	1115	25	25
Bean burrito	391	13	12	110	4	5	1138	19	20
Beef burrito	432	22	19	170	8	57	1303	16	22
Big beef burrito supreme	525	25	25	220	11	72	1418	20	25
Burrito supreme	443	18	19	170	9	47	1184	20	21
Chicken burrito	345	17	13	110	5	57	854	14	14
Chicken burrito supreme	520	27	23	200	9	125	1130	15	50
Chili cheese burrito	391	17	18	160	9	47	980	30	17
Combo burrito	412	17	16	140	6	32	1221	17	21
Steak burrito supreme	500	26	23	200	11	75	1350	20	15
Specialty Items									
Beef MexiMelt	262	13	14	130	7	38	711	23	8
Cinnamon twists	139	1	6	50	0	0	189	*	2
Mexican pizza	574	19	38	340	12	50	1003	31	25
Nachos	345	7	18	160	6	9	398	23	4
Nachos BellGrande	633	22	34	300	12	49	952	36	20
Nachos supreme	364	12	18	160	5	17	470	19	15
Pintos 'n cheese	190	9	9	80	4	14	640	15	7
Taco salad	838	31	55	490	16	79	1132	25	34
Side Orders & Condiments									
Green sauce	4	0	0	0	0	0	136	*	*
Guacamole	36	0	3	30	1	0	132	4	*
Hot taco sauce	2	0	0	0	0	0	91	*	*
Milk taco sauce	0	0	0	0	0	0	6	*	*
Nacho cheese sauce	51	2	4	40	2	4	196	6	*
Picante sauce	3	0	0	0	0	0	132	2	0
Pico de Gallo	6	0	0	0	0	0	65	*	*
Ranch dressing	136	1	14	130	3	20	330	2	*
Red sauce	10	0	0	0	0	0	261	*	*
Salsa	27	1	0	0	0	0	709	5	3
Seasoned rice	110	2	3	30	1	5	230	*	8
Sour cream	44	1	4	40	3	15	11	2	*

Source: Taco Bell Corporation, Irvine, CA. 1996.
* = trace

WENDY'S

Food Item	Serving Size (g)	kcal	Protein (g)	CHO (g)	Fat (g)	Choles-terol (mg)	Vit A (mg)	Vit C (mg)	Thia (mg)	Ribo (mg)	Nia (mg)	Ca (mg)	Fe (mg)
Sandwiches													
½ lb. hamburger patty	74	180	19	—	12	65	—	—	4	—	20	—	20
Plain single	126	340	24	30	15	65	—	—	25	20	30	10	30
Single with everything	210	420	25	35	21	70	5	15	25	20	30	10	30
Wendy's Big Classic	260	570	27	47	33	90	10	20	30	25	35	15	35
Jr. Hamburger	111	260	15	33	9	34	2	4	25	20	20	10	20
Jr. cheeseburger	125	310	18	33	13	34	2	4	25	50	20	10	20
Jr. bacon cheeseburger	155	430	22	32	25	50	2	15	30	50	25	10	20
Jr. Swiss deluxe	163	360	18	34	18	40	4	10	25	60	20	20	20
Kids' meal hamburger	104	260	15	32	9	35	2	2	25	20	20	10	20
Kids' meal cheeseburger	116	300	18	33	13	35	2	2	25	50	20	10	20
Grilled chicken fillet	70	100	18	—	3	55	—	—	4	4	35	—	6
Grilled chicken sandwich	175	340	24	36	13	60	2	8	30	25	50	10	20
Chicken breast fillet	99	220	21	11	10	55	—	—	8	8	60	—	70
Chicken sandwich	219	430	26	41	19	60	2	8	30	25	70	10	80
Chicken club sandwich	205	506	30	42	25	70	2	15	35	30	80	10	80
Fish fillet sandwich	170	460	18	42	25	55	2	2	40	35	20	10	15
Kaiser bun	65	200	6	37	3	10	—	—	25	20	10	10	10
White bun	56	160	5	30	3	tr	—	—	20	20	10	10	10
Sandwich Toppings													
American cheese slice	18	70	4	—	6	15	6	—	—	4	—	12	—
Bacon	6	30	2	—	3	5	—	4	4	—	2	—	—
Ketchup	14	17	—	4	—	NA	4	4	—	—	—	—	—
Lettuce	10	1	—	—	—	0	—	—	—	—	—	—	—
Mayonnaise	13	90	—	—	10	10	—	—	—	—	—	—	—
Mustard	5	4	—	—	—	0	—	—	—	—	—	—	—
Onion	10	4	—	—	—	0	—	—	—	—	—	—	—
Pickles	14	2	—	—	—	0	—	—	—	—	—	—	—
Tomatoes	21	4	—	—	—	0	—	6	—	—	—	—	—
Honey mustard	14	71	—	4	6	5	—	—	—	—	—	—	—
Tartar sauce	21	120	—	—	14	15	—	—	15	10	—	—	—
Superbar — Pasta													
Alfredo sauce	56	35	1	5	1	tr	—	—	—	—	—	6	—
Fettucini	56	190	4	27	3	10	—	—	10	6	6	—	6
Garlic toast	18.3	70	2	9	3	tr	4	—	6	2	2	2	2
Pasta medley	56	60	2	9	2	tr	6	15	6	4	4	—	4
Rotini	56	90	3	15	2	tr	—	—	6	4	6	—	4
Spaghetti sauce	56	28	—	7	0	tr	—	—	—	—	—	—	—
Spaghetti meat sauce	56	60	4	8	2	10	4	4	—	2	4	—	4
Garden Spot Salad Bar													
Alfalfa sprouts	28	8	1	1	0	0	—	4	—	2	—	—	—
Applesauce, chunky	28	22	—	6	—	0	—	—	—	—	—	—	—
Bacon bits	14	40	5	—	14	10	—	2	6	4	6	—	—
Bananas	28	26	—	7	—	0	—	4	—	2	—	—	—
Breadsticks	7.5	30	1	5	1	0	—	—	2	2	2	2	2
Broccoli	43	12	1	2	0	0	6	65	2	2	—	2	2
Cantaloupe	57	20	—	5	0	0	20	30	—	—	—	—	—
Carrots	27	12	—	2	0	0	80	4	2	—	—	—	—
Cauliflower	57	14	1	3	0	0	—	70	4	2	2	2	2
Cheddar chips	28	160	3	12	12	5	—	—	2	—	4	6	2
Cheese, shredded (imitation)	28	90	6	1	6	tr	4	—	—	15	—	20	—
Chicken salad	56	120	7	4	8	tr	—	4	—	4	6	—	2
Chives	28	71	6	18	1	0	195	313	15	25	8	25	30
Chow mein noodles	14	74	1	8	4	0	—	—	6	4	4	—	4
Cole slaw	57	70	—	8	5	5	4	25	—	—	—	2	—
Cottage cheese	105	108	13	3	4	15	6	—	—	10	—	6	—
Croutons	14	60	2	8	3	—	—	—	4	4	4	—	4
Cucumbers	14	2	—	—	0	0	—	—	—	—	—	—	—
Eggs (hard cooked)	20	30	3	—	2	90	4	—	—	6	—	—	—
Garbanzo beans	28	46	3	8	1	0	—	—	2	—	—	—	6
Green peas	28	21	1	4	0	0	4	8	6	2	—	—	2
Green peppers	37	10	—	2	0	0	2	60	2	—	—	—	—
Honeydew melon	57	20	—	5	0	0	—	25	2	—	2	—	—
Jalapeno peppers	14	2	—	—	0	0	—	—	—	—	—	—	—

WENDY'S—continued

Food Item	Serving Size (g)	kcal	Protein (g)	CHO (g)	Fat (g)	Choles- terol (mg)	Percentage of U.S. RDA						
							Vit A (mg)	Vit C (mg)	Thia (mg)	Ribo (mg)	Nia (mg)	Ca (mg)	Fe (mg)
Garden Spot Salad Bar—continued													
Lettuce—iceberg	55	8	—	1	0	0	2	4	2	—	—	—	2
Lettuce—romaine	55	9	1	1	0	0	15	20	4	4	—	2	4
Mushrooms	17	4	—	—	0	0	—	—	—	4	4	—	—
Olives, black	28	35	—	2	3	0	—	—	—	—	—	2	4
Oranges	56	26	—	7	0	0	—	50	4	—	—	2	—
Parmesan cheese	28	130	12	1	9	20	6	—	—	6	—	40	—
Parmesan cheese (imitation)	28	80	9	4	3	tr	20	—	—	—	—	50	—
Pasta salad	57	35	2	6	—	0	—	—	—	2	2	—	2
Peaches	57	31	—	8	0	0	2	2	—	—	2	—	—
Pepperoni, sliced	28	140	5	2	12	35	—	—	160	4	10	—	2
Pineapple chunks	100	60	—	16	0	0	—	15	6	—	—	—	2
Potato salad	57	125	—	6	11	10	—	10	2	—	2	—	2
Pudding—butterscotch	57	90	1	11	4	tr	—	—	—	—	—	6	2
Pudding—chocolate	57	90	—	12	4	tr	—	—	2	2	—	15	2
Red onions	9	2	—	—	0	0	—	—	—	—	—	—	—
Red peppers, crushed	28	120	5	15	4	0	200	15	10	15	20	2	15
Seafood salad	56	110	4	7	7	tr	—	2	—	2	—	20	2
Strawberries	56	17	—	4	0	0	—	50	—	2	—	—	—
Sour topping	28	58	—	2	5	0	—	—	—	—	—	—	—
Sunflower seeds & raisins	28	140	5	6	10	0	—	—	30	4	6	2	10
Three bean salad	57	60	1	13	—	—	4	—	—	—	—	—	2
Tomatoes	28	6	—	1	0	0	2	10	—	—	—	—	—
Tuna salad	56	100	8	4	6	tr	—	4	—	4	25	—	2
Turkey ham	28	35	5	—	1	15	—	—	—	4	6	—	4
Watermelon	57	18	—	4	0	0	2	10	4	—	—	—	—
Salad Dressings													
Blue cheese	15	9	—	—	10	10	—	—	—	—	—	—	—
Celery seed	15	70	—	3	6	5	—	—	—	—	—	—	—
French	15	60	—	4	6	0	—	—	—	—	—	—	—
French, sweet red	15	70	—	5	6	0	—	—	—	—	—	—	—
Hidden Valley ranch	15	50	—	—	6	5	—	—	—	—	—	—	—
Italian Caesar	15	80	—	—	9	5	—	—	—	—	—	—	—
Italian, golden	15	45	—	3	4	0	—	—	—	—	—	—	—
Salad oil	28	250	0	0	28	0	—	—	—	—	—	—	—
Thousand island	15	70	—	2	7	5	—	—	—	—	—	—	—
Wine, vinegar	15	2	—	—	0	0	—	—	—	—	—	—	—
Reduced Calorie bacon & tomato	15	45	—	3	4	—	—	—	—	—	—	—	—
Reduced calorie Italian	15	25	—	2	2	0	—	—	—	—	—	—	—
Prepared Salads													
Chef salad	331	180	15	10	9	120	110	110	15	25	6	25	15
Garden salad	277	102	7	9	5	0	110	110	10	20	6	20	10
Taco salad	791	660	40	46	37	35	80	80	30	45	25	80	35
Superbar—Mexican Fiesta (where available)													
Cheese sauce	56	39	1	5	2	tr	—	—	—	—	—	6	—
Picante sauce	56	18	—	4	—	NA	10	30	2	—	2	—	2
Refried beans	56	70	4	10	3	tr	—	—	4	2	—	2	6
Rice, Spanish	56	70	2	13	1	tr	6	—	45	—	8	4	10
Taco chips	40	260	4	40	10	0	—	—	2	4	—	8	4
Taco meat	56	110	10	4	7	25	—	—	8	6	10	4	10
Taco sauce	28	16	—	3	—	tr	4	2	—	—	—	—	—
Taco shells	11	45	—	6	3	0	—	—	—	—	—	—	—
Tortilla, flour	37	110	3	19	3	NA	—	—	4	2	2	8	2
French fries (small) 3.2 oz**	91	240	3	33	12	0	—	10	10	2	10	—	4
Chili (regular) 9 oz	255	220	21	23	7	45	15	15	8	10	10	8	35
Cheddar cheese, shredded	28	110	7	1	10	30	10	—	—	6	—	20	—
Sour cream	28	60	1	1	6	10	6	—	—	2	—	4	—
Crispy chicken nuggets (6)	93	280	14	12	20	50	—	—	6	6	30	4	4

WENDY'S — continued

Food Item	Serving Size (g)	kcal	Protein (g)	CHO (g)	Fat (g)	Choles-terol (mg)	Percentage of U.S. RDA						
							Vit A (mg)	Vit C (mg)	Thia (mg)	Ribo (mg)	Nia (mg)	Ca (mg)	Fe (mg)
Nuggets sauces													
Barbeque	28	50	—	11	—	0	6	—	—	—	—	—	4
Honey	14	45	—	12	—	0	—	—	—	—	—	—	—
Sweet & sour	28	45	—	11	—	0	—	—	—	—	—	—	2
Sweet mustard	28	50	—	9	1	0	—	—	—	—	—	—	—
Hot Stuffed Baked Potatoes													
Plain	250	270	6	63	—	0	—	50	20	6	20	2	20
Bacon & cheese	362	520	20	70	18	20	10	60	35	15	35	8	24
Broccoli & cheese	350	400	8	58	16	tr	14	60	20	10	20	10	15
Cheese	318	420	8	66	15	10	10	50	20	100	20	6	20
Chili & cheese	403	500	15	71	18	25	15	60	20	100	25	8	28
Sour cream & chives	323	500	8	67	23	25	50	75	20	10	20	10	20
Beverages													
Frosty, small***	243	400	8	59	14	50	10	—	8	30	2	30	6
Cola, small	8*	100	0	25	0	0	—	—	—	—	—	—	—
Lemon-lime soft drink, small	8*	100	0	24	0	0	—	—	—	—	—	—	—
Diet cola	8*	1	0	—	0	0	—	—	—	—	—	—	—
Coffee	6*	2	0	—	0	0	—	—	—	—	—	—	—
Decaf coffee	6*	2	0	—	0	0	—	—	—	—	—	—	—
Hot chocolate	6*	110	2	22	1	tr	—	—	—	8	—	6	2
Lemonade	8*	90	0	24	0	0	—	15	—	4	—	—	2
Choc milk	8*	160	7	24	5	15	15	4	6	20	—	25	4
Milk, 2%	8*	110	8	11	4	20	10	4	6	20	—	30	—
Tea (hot or ice)	6*	1	0	0	0	0	—	—	—	—	—	—	—
Chocolate chip cookie	64	275	3	40	13	15	2	—	8	8	6	2	8

Source: Consumer Relations, Wendy's International, Dublin, OH.
*Fluid ounces.
**To determine nutritional information for a large order of Fries, multiply figures by 1.3; Biggie Fries, multiply by 1.87; large Chili, multiply by 1.5; 9-piece Nuggets, multiply by 1.5; 20-piece Nuggets, multiply by 3.3.
***To determine nutritional information for a medium Frosty, multiply figures by 1.3; large Frosty, multiply by 1.7. For medium soft drink, multiply by 1.5; large soft drink, multiply by 2. For Biggie soft drink, multiply by 3.5.